# HOW IT WORKS

THE CALIBER .50. M2 BROWNING MACHINE GUN

HOW IT WORKS

The Caliber .50. M2 Browning Machine Gun

Copyright© 2021 ISBN: 978-1-955611-99-2

This book provides reprints of the now very collectable WW II "Training Manual for the Caliber .50, M2 Browning Machine Gun, Aircraft Basic" and the "How The Gun Works" pamphlet for the same gun. Both publications were produced by the AC Spark Plug Division, General Motors Corporation in conjunction with Frigidaire Division, General Motors Corporation. Published in 1943.

Cover Photo Credit:

**FORT MCCOY, WI, UNITED STATES 07.20.2014**

**Photo by** Staff Sgt. Chad Nelson 135th Mobile Public Affairs Detachment

**Formatted by** HMG Formatting, hmgformatting@gmail.com, Kalispell MT.

Z-Hat Publishing, Kalispell MT

# A Short History of the
# Browning 50 Caliber Machine Gun

By Fred Zeglin

You have probably already guessed from the name that this gun derived from the genius of non-other than John Moses Browning. He was and is the most prolific American gun inventor.

Browning designed guns of all sizes and types. His contributions the military firearms are amazing and include: The M1895 Machine Gun, the 1911 pistol, the Browning 1917/1919, the BAR, the Browning Hi- Power, the 37mm M4 Auto Cannon and the M2 (Ma Deuce) Browning Machine gun which is our primary subject here.

During World War I General John J. Pershing requested a new design for use against armored vehicles as the Germans had begun to put heavy armor on many of their vehicles.

*Figure 1 General John J. Pershing*

Initial attempts to meet General Pershing's demand for a heavy machine gun started with Col. John Henry Parker taking notice of the French 11x59mm Gras- Vickers loaded with incendiary and armor piercing bullets used by the French to knock down enemy observation balloons. Col. Parker detailed his subordinate, Capt. Henry B. Allen to look into secure as much information as possible about the machine gun and the ammunition.

Captain Allen was able to secure a test gun to be shipped to the U.S. for inspection and testing. The problem of course is the 11mm cartridge was not even close to meeting the demands of the needs laid out by Pershing. It had a 385 grain bullet with a velocity of 1490 feet per second. After a few failed attempts by government engineers the problem was taken to Browning who at the time was overseeing production at Colt.

*Figure 2 Col. John Henry Parker*

In about July of 1917, Browning told the government representative jokingly, "You load up some ammo and we will do some shooting." As it turned out the Browning prototype was ready before the ammo.

Winchester undertook the job of cartridge design with Browning consulting. At first Winchester offered the design with a rimmed cartridge for use both in the machine gun and in a proposed clip-fed, Bolt action anti-tank rifle that Winchester was making at the time. See this episode of Forgotten Weapon on YouTube: https://voutu.be/FchMnsd3YSk "Prototype Winchester WW1 .50 Cal Antitank Rifle".

When specification for the Winchester designed round crossed Pershing's desk he cabled the Ordnance Department on 18 July, 1918 rejecting the design and ordering that it be redesigned as a rimless cartridge. The resulting 50 BMG round is pretty much a scaled up 30-06, so everything is much larger. Pershing had laid out requirements for the new cartridge, a projectile of at least 670 grains and a velocity of 2700 feet per second (fps).

Working in conjunction with Colt the first working model of the 50 caliber Browning was completed when the parts were delivered to Browning at the Winchester plant on September 12, 1918. Browning had moved is prototype gun to Winchester to speed the development of the ammunition, firing his gun as a single shot.

At that point Winchester agreed to build six more guns to fully test and debug the new design. On October 15, 1918 the first 50 BMG machine gun was ready for the proving ground.

It was too late for use in the First World War, but four days after the armistice was signed on November 11, 1918, the gun received its government trial at the Aberdeen Proving Ground. John Browning personally fired the gun without a single failure or malfunction. 877 rounds went down range in burst of about 100 to 150 rounds each. Rate of fire was 500 rounds per minute. These first loadings had a 707 grain bullet at 2300 fps, falling short of the desired velocity set forth by General Pershing.

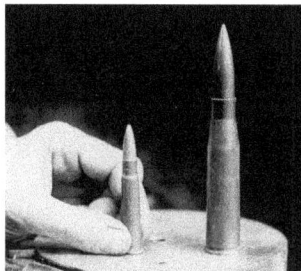

Figure 3 303 British vs. the 13.2mm TuF

Impressed by the test, the Engineering Division of the Office of Chief of Ordinance recommended that Winchester be given an order for 10,000 guns. The first model was water cooled had a 30.5" barrel, the longest that Winchester could make at the time. Winchester promised to balance the loading and provide longer barrels for the finished design in order to increase the velocity.

Before development had moved too far along, a few German anti-tank rifles were *Figure 3 303 British vs. the 13.2mm TuF*captured, we know them today as the Mauser Tankgewehr, together with some of the 13.2mm TuF ammunition they fired. Ballistics of the 13.2mm TuF: the Germans were getting a muzzle velocity of 2575 fps with a 795 grain bullet. The round would easily penetrate the armor of the WW I British MK I Tanks.

*Figure 4 Two Officers with a captured Mauser Tankgewehr*

Winchester and the Ordnance Department recognized that adjustments to their project were necessary or it would be obsolete before it got to the field. So Winchester immediately undertook the process of exceeding the German antitank rifle cartridge ballistics.

On December 1, 1918 the Ordnance Department decided that all future development of the caliber 50 ammunition would be done at Frankford Arsenal, instead of at the Winchester plant. This transition took some time as Frankford Arsenal had to tool up and take over production.

When designing the M2 Browning took inspiration from his own 1917 machine gun design to scale up for the 50 caliber. Two major changes were the spade grips and the addition of the oil buffer which offered two benefits to the design. First, it absorbed excess energy from recoiling parts resulting from the increased powder charge. Second, it provided a method of regulating rate of fire.

The "Ma Deuce" as developed by John Browning was adopted for service in 1921 in a water cooled version of the design: the M1921. Which allowed for a higher cyclic rate and was seen as an anti-aircraft gun. There are many variations of the M2. The gun we're looking at in this book is the AN M2. Which stands for Army/Navy M2, this gun is lighter and utilizes a full barrel jacket over a lighter weight barrel. Ground units would be using the M2 HB or Heavy Barrel.

*Figure 5 John M. Browning and Mr. Burton
studying a BAR at the Winchester Repeating
Arms Company plant, Connecticut, United States
circa 1918*

The total time consumed by John M. Browning on the Caliber .50 machine gun from conception to successful firing was slightly over a year. When asked by the press to what he attributed his achievement, he replied, "One drop of genius in a barrel of sweat wrought the miracle."

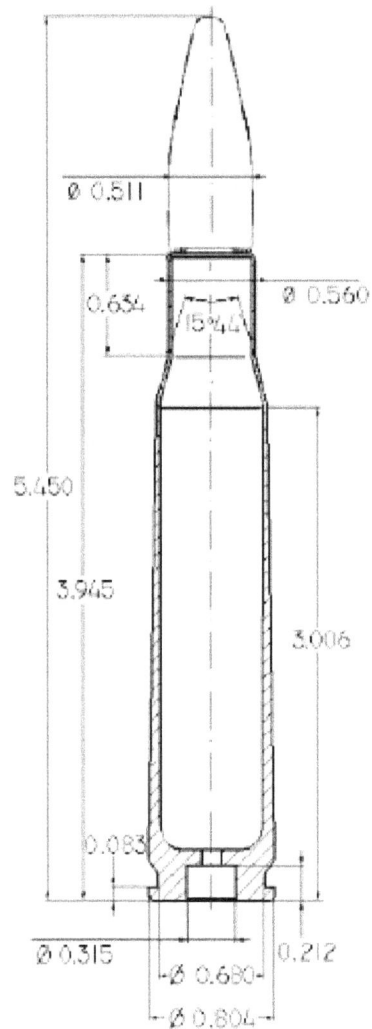

*Figure 6 .50 BMG aka, 12.7x99mm NATO*

Browning read some years later that Thomas Edison was quoted as saying, "Genius is one percent inspiration and ninety-nine percent perspiration." Browning decided that Edison had put it better, "but it was still just plain old sweat."

One of my favorite stories about John Browning comes from his biography and concerns his compensation for the guns he designed from the U.S. Government. An officer was sent to meet with Browning and make an offer for the rights to produce his designs; the 1917 Machine gun, the BAR and the 1911 pistol. John's brother, Matt was present and related the conversation.

The officer stated, "The offer, we know, is only a fraction of what you would receive from royalties on orders already booked, and it may not be acceptable. In that event negotiations will be necessary." He then delivered the offer.

Matt tells of the meeting, "I supposed that John would ask for a little time to think things over and get my opinion, but without hesitating a second, he said, 'Major, if that suits Uncle Sam, it's all right with me.'"

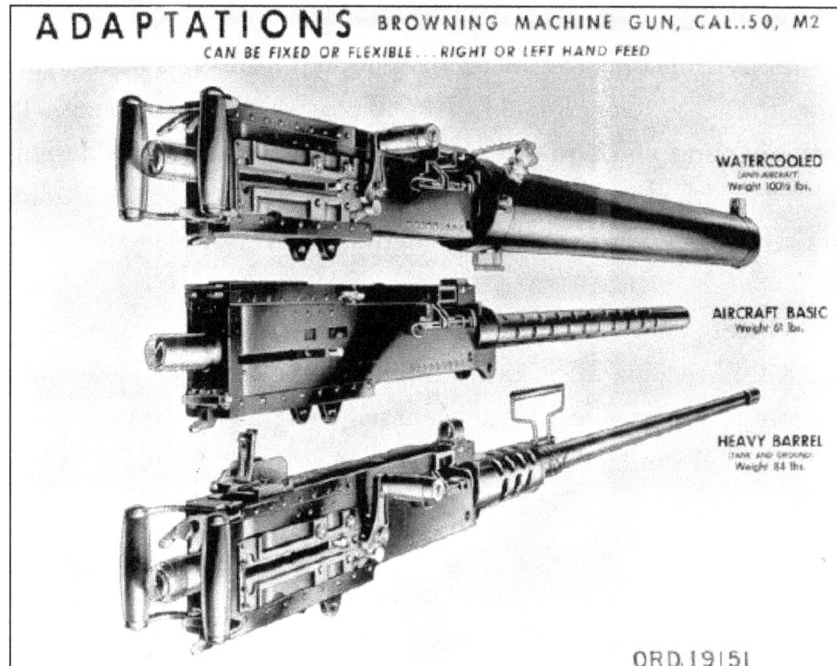

ADAPTATIONS BROWNING MACHINE GUN, CAL..50, M2
CAN BE FIXED OR FLEXIBLE...RIGHT OR LEFT HAND FEED

WATERCOOLED
(ANTI-AIRCRAFT)
Weight 100½ lbs.

AIRCRAFT BASIC
Weight 61 lbs.

HEAVY BARREL
(TANK AND GROUND)
Weight 84 lbs.

ORD. 19151

After the officer departed Matt asked is brother why not negotiate we could have received much more without any real argument. John answered his brother, "Yes, and if we were fifteen or twenty years younger we'd be over there in the mud!"

Following the death of Browning in 1926 his design was modified for use on aircraft and on the ground.

The M2 HB weighs 84 lbs. by itself, if you add a tri-pod the weight is 128 lbs. Almost two million of these guns were produced for the war effort in WW II. The design has been adapted for many jobs by the militaxy. It is the longest serving small arm in the American Armed Forces, it is still in wide service even as of this writing.

Cyclic rates vary depending on the adaptation and its intended use. The AN M2 has a rate of fire of about 850 rounds per minute. This version weighs in at a svelte 61 lbs. and became the gun of choice for American aircraft.

*Figure 7 Side View, M2 Heavy Barrel*

*Figure 8   M2 HB*

Figure 9 M2 Nose Gun, 41st Bombardment
Group - Hawkins Field (Tarawa)

*Figure 10 Twin Mounted 50s*

For more reading on the subjects covered here see:
The Machine Gun by Georg M. Chinn, Lieutenant Colonel,
USMC, Published by the Bureau of Ordnance, Department
of the Navy.
And

John M. Bowning, American Gunmaker by John Browning and
Curt Gentry, Library of Congress # 64-19275

Figure 11 Fred Zeglin, editor of this short history
of the Ma Deuce, trying one out.

# TRAINING MANUAL

## CALIBER .50, M2
### BROWNING MACHINE GUN
### AIRCRAFT BASIC

★

THIS MANUAL IS
IDENTIFIED AS
**FGA**

Prepared by—

AC SPARK PLUG DIVISION, General Motors Corporation
FLINT, MICHIGAN, U. S. A.

FRIGIDAIRE DIVISION, General Motors Corporation
DAYTON, OHIO, U. S. A.

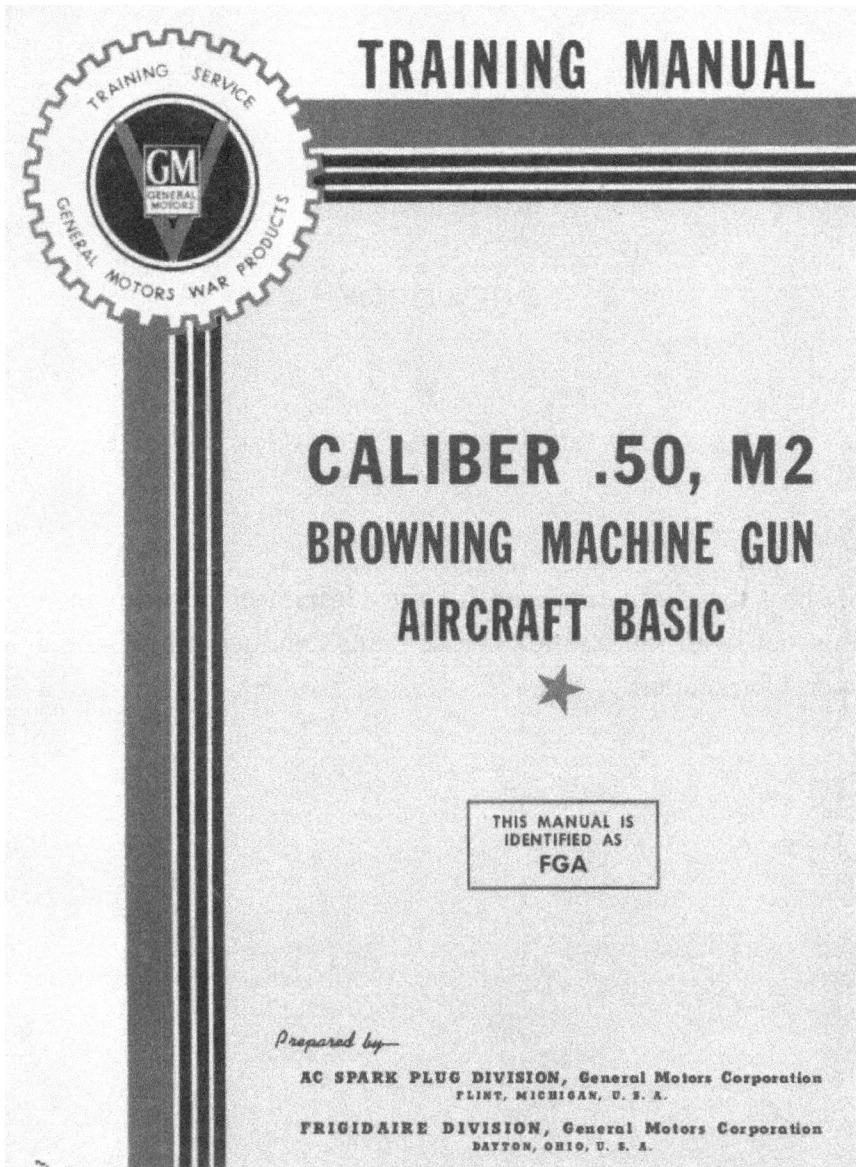

*Format of original document was 8.5"x11". Reprinted hereafter in its entirety.*

# FOREWORD

This book has been prepared for the Ordnance Department in connection with the training course being conducted by General Motors Corporation.

# INTRODUCTION

The Browning Machine Gun, caliber .50, M2, aircraft basic, is a highly efficient automatic weapon built to precision standards. It is an air-cooled, alternate feed gun and, as such, may be fed from either the right or left side. It may be mounted on either a rigid or flexible mount. Manual cocking and loading of the gun is necessary before it is ready to -function automatically. Then it may be fired by a mechanical or electrical accessory or by a manual trigger and trigger bar. While firing, all mechanical action is automatically performed by the gun itself and it will fire between 750 and 850 rounds per minute.

To care for the gun properly, so it will fire and keep on firing when needed, it is essential that its mechanical action be thoroughly understood and this manual is published for that purpose.

The general appearance of this gun may be noted in Figure 1, which shows a top and side view of the complete aircraft basic gun.

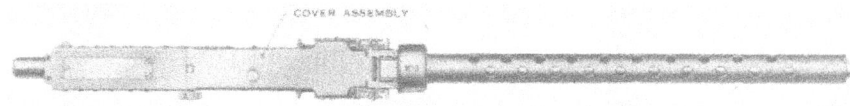

Top View

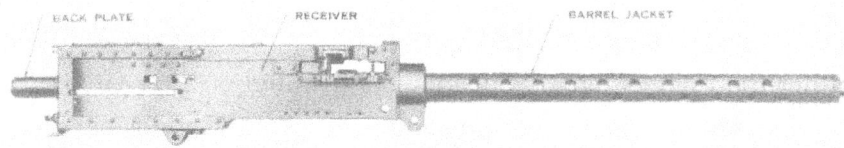

Side View

*Figure 1 Browning Machine Gun Caliber .50, M2, Aircraft Basic*

## RECEIVER AND BARREL JACKET GROUP

The receiver and barrel jacket form the main exterior portions of the gun, and in gun operation are stationary or non-recoiling. These separate assemblies may be noted in Figure 2 and their relative position in the gun maybe seen in Figure 1.

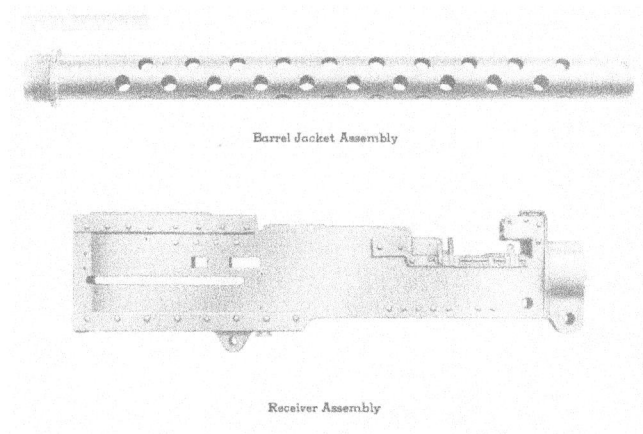

Barrel Jacket Assembly

Receiver Assembly

Figure 2 Receiver and Barrel Jacket Group

## BACK PLATE GROUP

The back plate is assembled to the rear of the receiver, forming an end cover. See Figure 3 and Figure 1.

Right Side View

Bottom View

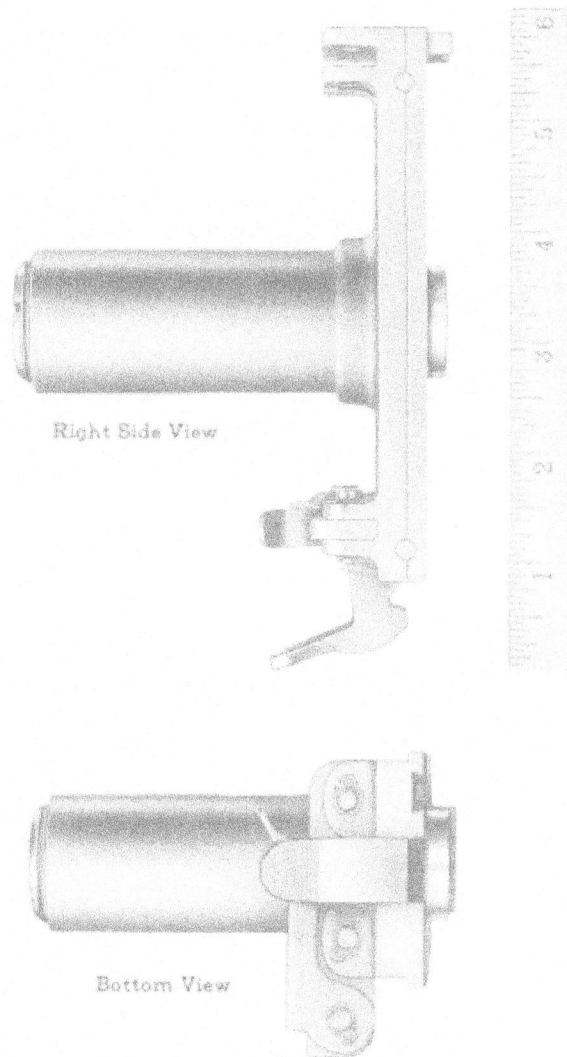

Figure 3 Back Plate Group

BOLT GROUP (alternate feed)

The driving spring runs lengthwise of the gun with one end resting 'against the back plate. The bolt group is housed in the upper forward portion of the receiver, and slides backward and forward during operation. See Figure 4.

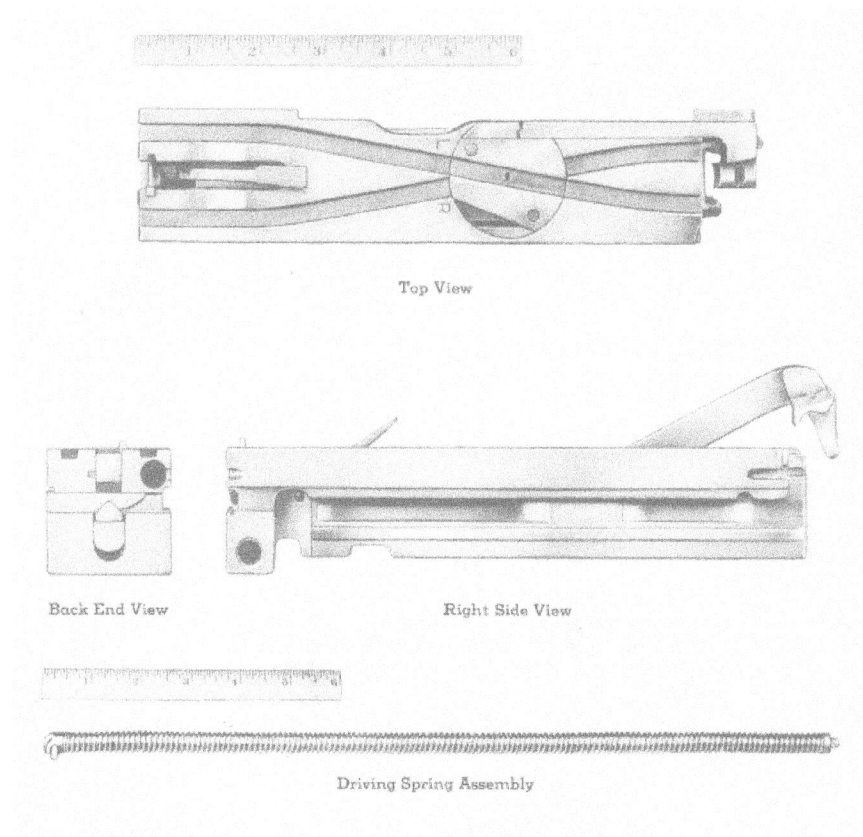

Top View

Back End View                    Right Side View

Driving Spring Assembly

Figure 4 Bolt Group (Alternate Feed) and Driving Spring

## OIL BUFFER BODY AND OIL BUFFER GROUP

The oil buffer body and oil buffer group are contained in the lower rear portion of the receiver. The bolt slides over the top of the oil buffer body during the back portion of the stroke. See Figure 5.

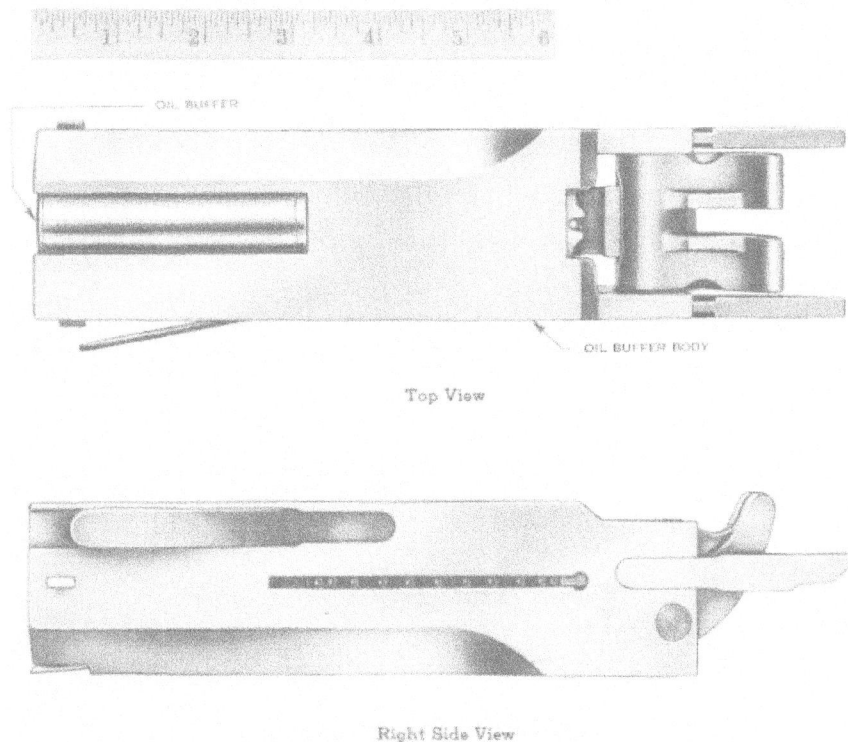

Figure 5 Oil Buffer Body and Oil Buffer Group

## BARREL AND BARREL EXTENSION GROUP

The barrel extension is screwed onto the breech end of the barrel to form a single unit, and the barrel slides inside of the barrel jacket. Thus the normal position of the barrel extension is in the lower forward portion of the receiver. See Figure 7. The bolt slides in grooves of the barrel extension.

Figure 6 Barrel

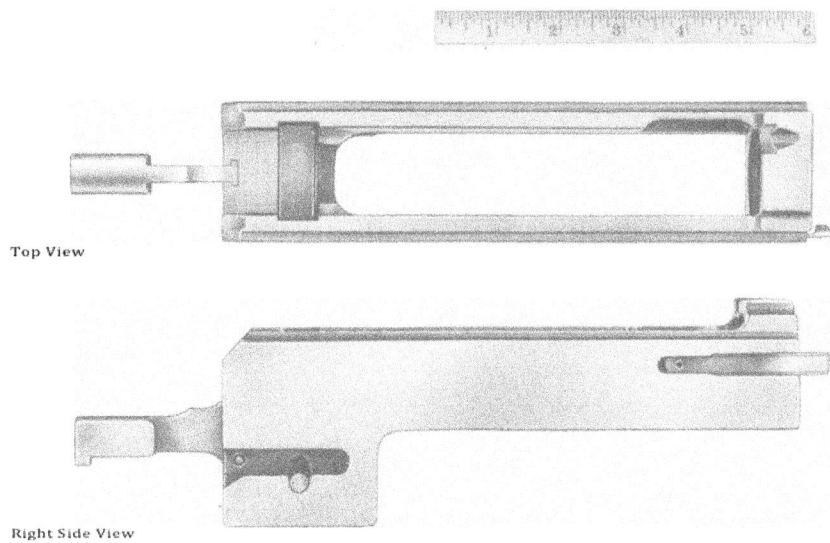

Top View

Right Side View

Figure 7 Barrel Extension Group

## COVER AND BELT FEED GROUP

The cover and belt feed group is hinged at its forward end to the top front portion of the receiver. See Figure 8 and Figure 1.

COVER AND BELT FEED GROUP

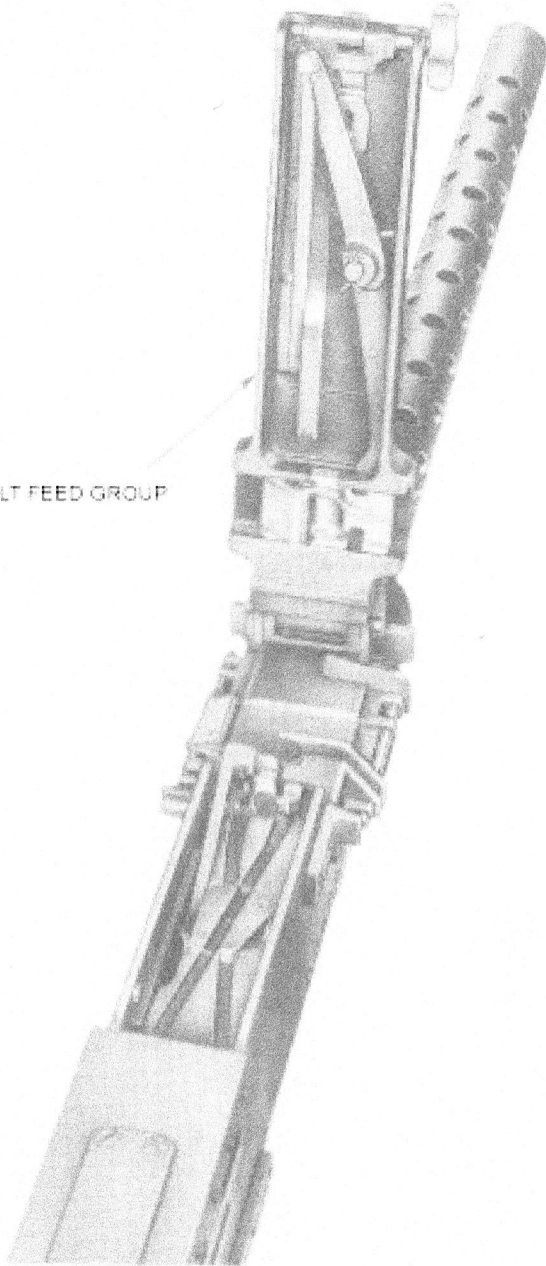

# SPECIFICATIONS

## BROWNING MACHINE GUN, CALIBER .50, M2
### Aircraft Basic

GENERAL DATA (approximate)

| | |
|---|---|
| Weight of Gun .................. ........................... | 64 lbs. |
| Weight of Bullet ......................................... | 750 grains (1.71 oz.) |
| Weight of Powder Charge........................ | 200 grains (0.46 oz.) |
| Weight of Cartridge ................................... | 4.20 oz. |
| Weight of 100 Cartridges in Metallic | 30.25 lbs. |
| Links | 10 lbs. |
| Weight of Barrel ........................................ | 36 inches 8 |
| Length of Barrel ........................................ | 1 turn in 15 inches |
|     Number of Lands.................................... | 56.125 inches |
|     Twist—Right Hand............................... | . 2,750 ft. per sec. (1,875 mi. per hr.) |
| Overall Length of Gun .............................. Muzzle Velocity......................................... | . 750 to 850 rounds per minute |
| Rate of Fire............................................... | .7,200 yards (4.1 miles) |
| Maximum Range........................................... | |

Figure 8A Ammunition Belt Links

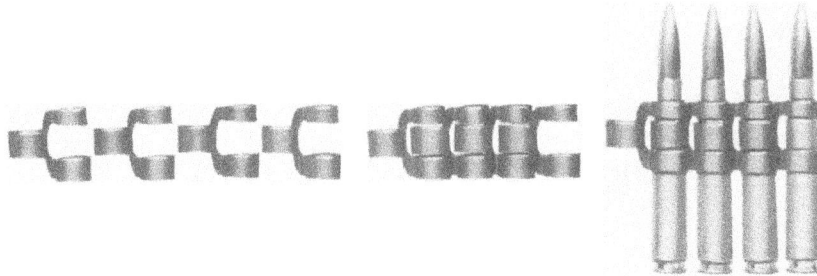

Single cartridges or rounds are first assembled into a series of nested links as shown in Figure 8A. This forms a flexible ammunition belt with a single unused link trailing on one end and a double used link on the other. The double link end is fed into the gun.

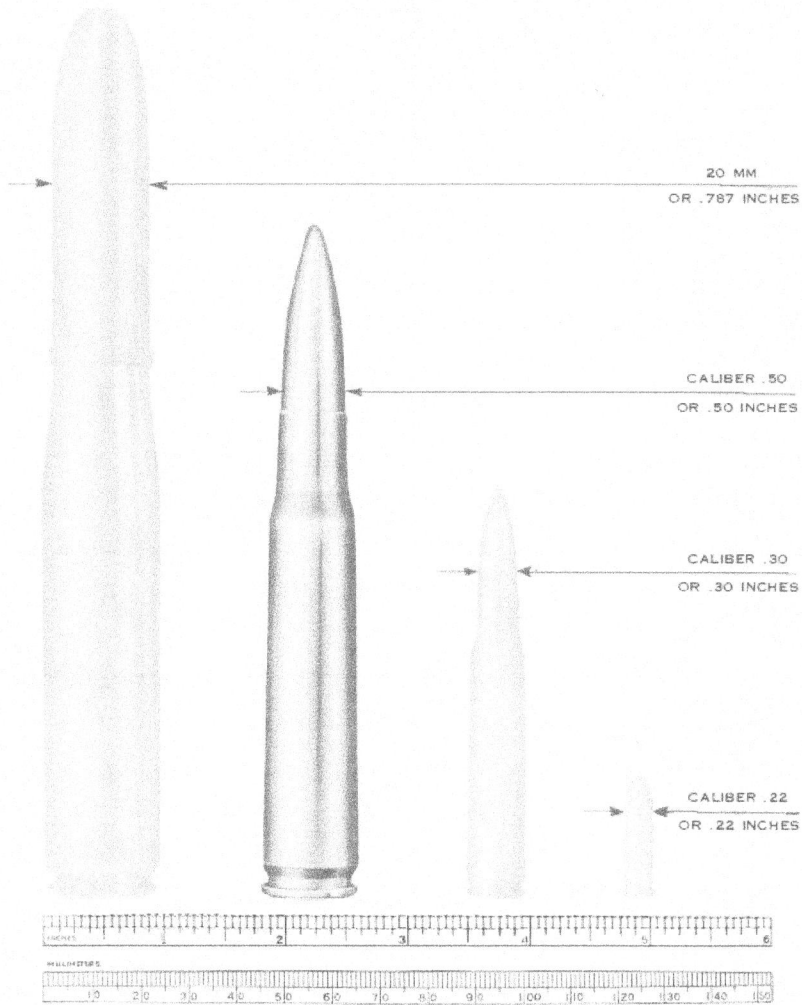

20 MM
OR .787 INCHES

CALIBER .50
OR .50 INCHES

CALIBER .30
OR .30 INCHES

CALIBER .22
OR .22 INCHES

Figure 9 Caliber Illustration

# GENERAL FUNCTIONING

The Browning Machine Gun, Caliber .50, M2, is an air-cooled gun capable of attaining a rate of fire between 750 and 850 rounds a minute.

Although this gun is an automatic weapon, it is necessary to "cock" it manually to start the operating sequence. Assume that the gun is cocked, and that the first cartridge is in its firing position in a chamber in the rear portion of the gun barrel.

When any cartridge is fired, the burning powder violently generates gas which, since it is confined by the cartridge case and barrel, exerts a tremendous pressure. This pressure reaches 50,000 pounds per square inch, and since this pressure pushes against the back face of the bullet, which up to this moment is still within the cartridge case and which has a diameter of one-half inch, a driving force of five tons pushes the bullet out of the barrel. **This same force tries to drive the cartridge case out of the**

**chamber toward the rear.** Such action is prevented by having the bolt positively locked against the rear of the cartridge at the instant of firing.

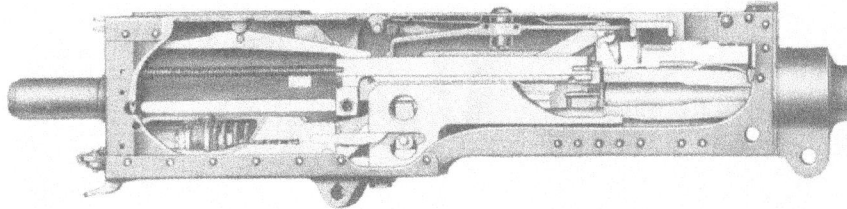

Figure 9A Cutaway View of Gun

When the cartridge is fired the force of recoil carries the barrel, barrel extension, and bolt (known as the recoiling portion) backward a short distance. See Figure 10A. This motion unlocks the bolt from the barrel and barrel extension, and the bolt throws back farther toward the rear against a spring. This spring serves to drive the bolt forward again. The empty case is withdrawn by the bolt from the barrel chamber and the next cartridge is extracted from the supply belt.

The long rearward motion ot the bolt is checked and as it surges forward the empty case is ejected and the next cartridge is moved into the barrel chamber. See Figure 10B. The short, rearward motion of the barrel and barrel extension is checked by the oil buffer and its spring; this buffer spring serves to drive them forward again. This motion locks the bolt to the barrel, thus again preventing the case from being driven toward the rear. The last forward motion of the bolt and barrel causes the firing pin to strike the cartridge, providing a means of releasing the sear is furnished.

This cycle continues as long as trigger action is maintained and as long as ammunition is supplied.

## Recoiling Portion

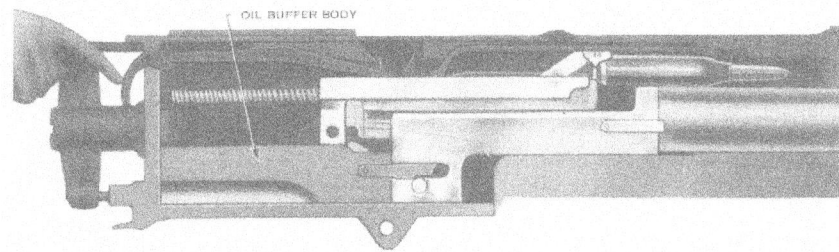

Figure 10A

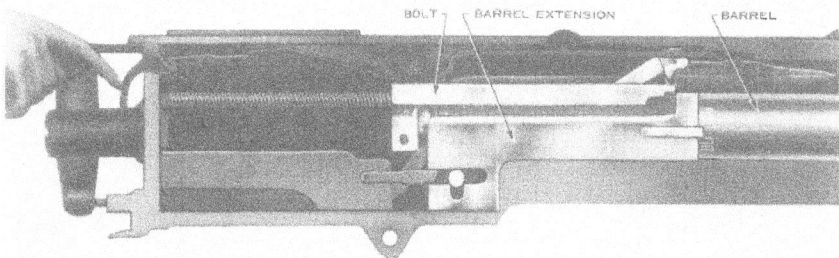

Figure 10B

## BACK PLATE GROUP

The back plate, besides enclosing the back end of the receiver, also houses the final bolt recoil cushioning parts. See Figure 11. It also acts as a stop for the oil buffer group. The plate slides downward into grooves in the receiver side plates and is latched in place with a latch and latch lock.

Assembled and compressed into the projecting cylindrical portion is a stack of lightly greased fiber discs which are headed on the forward end by the buffer plate. The final movement of the bolt is stopped by the fiber discs as the bolt strikes this buffer plate, and these discs assist in starting the forward movement.

Although the back end of the driving spring group is retained during assembly by inserting the driving spring rod retaining pin into a hole in the receiver, in actual operation the force of the spring is counteracted by the back plate.

The back plate furnished with the basic gun is intended for fixed mount, remote firing applications. If the gun is to be used in a flexible manner, the necessary handle and manual trigger parts are added. An auxiliary filler piece is supplied with the basic gun to close the opening provided for a manual trigger.

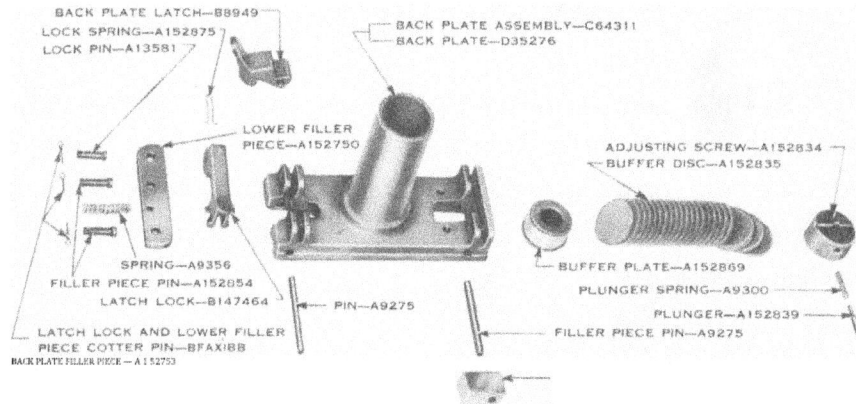

Figure 11 Bach Plate Details

## BOLT GROUP (alternate feed)

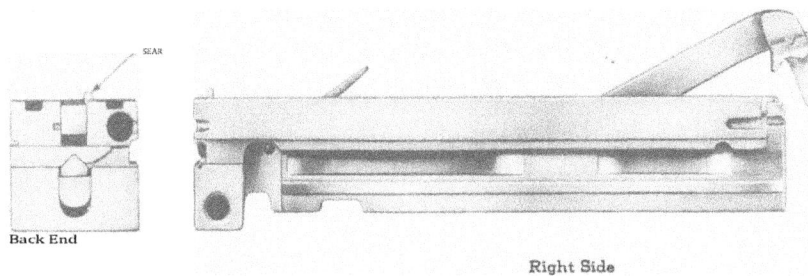

Figure 12 Bolt Group (alternate feed)

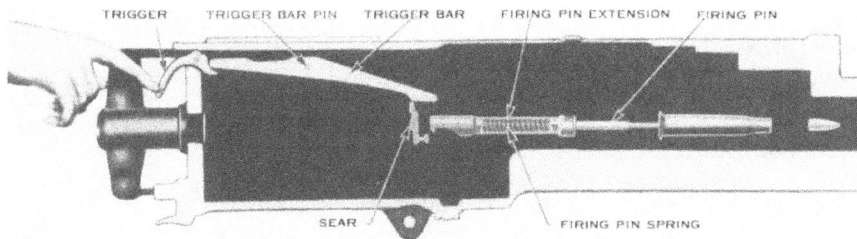

Figure 13 Firing Mechanism

The bolt group is probably second in importance to the receiver. It holds the cartridge firmly in the chamber when it is fired; it withdraws the empty case and ejects; it extracts a fresh cartridge from the belt and inserts it in the chamber; it actuates the belt feed mechanism. The sear mechanism, when actuated by a trigger, trips the cocked firing pin, causing the gun to fire.

As may be noted in Figure 12, the sear, which moves vertically in the back end of the bolt, may be depressed by pushing down on the small protrusion which extends up beyond the top of the bolt. When a manual trigger is supplied, the sear is depressed in this manner by means of the trigger bar in the top of the receiver. The downward motion of the sear unhooks or releases the firing pin extension, Figure 13. This extension, along with the firing pin, snaps forward under the action of the cocked or compressed firing pin spring, and the tip of the firing pin protrudes from the front end of the bolt, thus striking the cartridge primer.

BOLT GROUP (alternate feed)

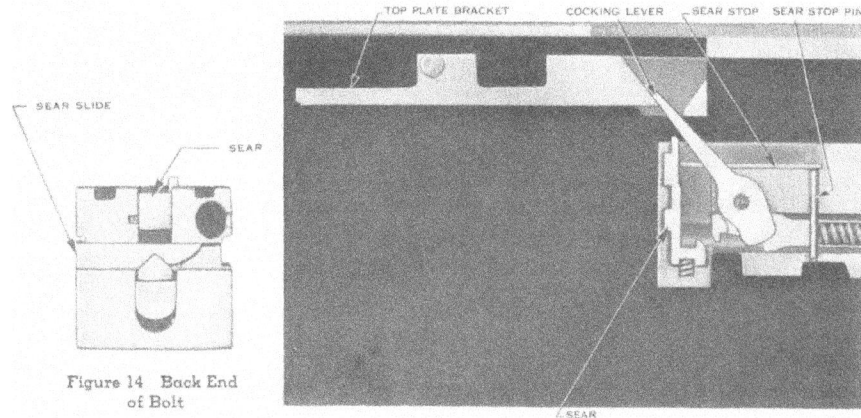

Figure 14  Back End of Bolt

Figure 15 Cocking Mechanism

The sear may also be actuated by side pressure on the end of the sear slide, Figure 14. This slide may be assembled either right or left hand, and suitable openings in both side plates of the receiver permit the gun to be fired from either the right or the left side. The necessary pressure may be supplied by electrical (solenoid) means or by a cable attachment which fastens to the side plate.

Although the sear is constantly being forced upward by the sear spring, it is retained in its slot by the sear slide. See Figure 14. The sear stop pin projects downward through the body of the bolt to act as a stop for the firing pin spring. See Figure 15.

The cocking lever, which at its lower end engages with a slot in the firing pin extension, has its top end projecting above the bolt. This top end engages with a cam in the top plate bracket of the receiver to cock the firing pin.

BOLT GROUP (alternate feed)

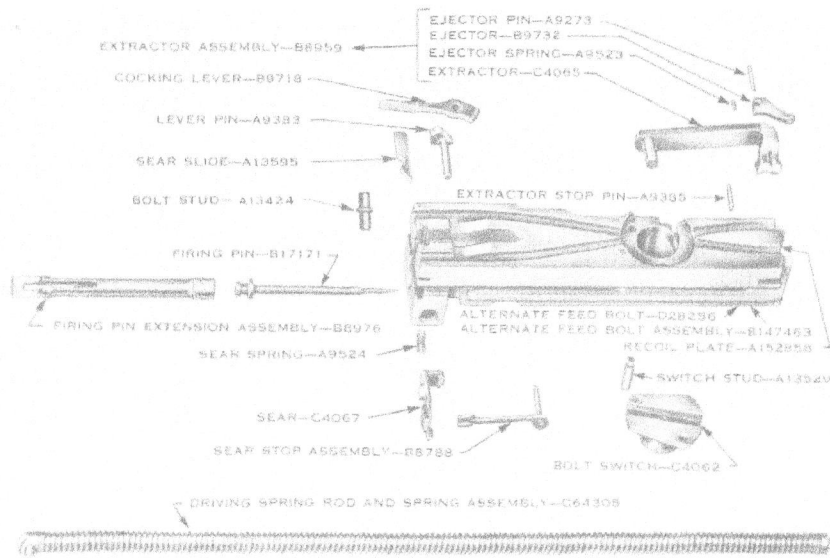

Figure 16 Bolt Group Details

The top surface of the bolt has two diagonal ways or grooves which act as cams to actuate the belt feed mechanism in the cover. See Figure 16. The bolt switch fits into the circular depression on top of the bolt and may be assembled to make one or the other of these two ways continuous, the selection depending on whether ammunition will be fed from the right or left side.

The extractor which fits into a circular hole on the left-hand side of the bolt, withdraws a cartridge from the belt and places it in the T-slot at the front end of the bolt. The extractor stop pin stops the extractor in its downward swing on the counter-recoil stroke.

The ejector is fastened to the end of the extractor, and helps to position a new cartridge in the feed-way when ammunition is being fed from the right-hand side. It also guides a new cartridge into the chamber, and pushes the last empty case out of the T-slot.

A driving spring assembly fits into a lengthwise hole in the bolt, and is compressed by the rearward motion of the bolt. After the bolt recoil has been stopped by the back plate, the spring drives the bolt forward. This assembly actually has two springs, one nested inside the other, and both are slipped over a rod with suitable end retaining parts.

## BARREL AND BARREL EXTENSION GROUP

BARREL—D35348A

Figure 17 Barrel

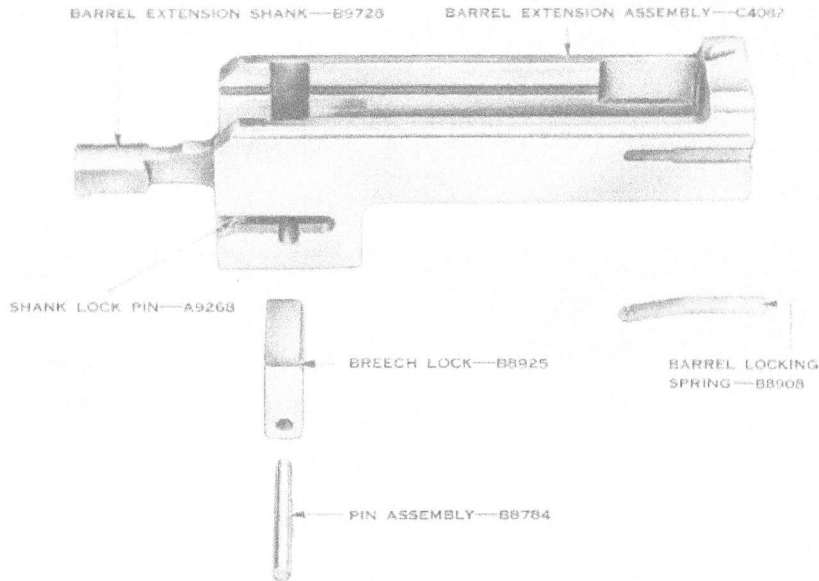

BARREL EXTENSION SHANK—B9728      BARREL EXTENSION ASSEMBLY—C4082

SHANK LOCK PIN—A9268

BREECH LOCK—B8925

BARREL LOCKING
SPRING—B8908

PIN ASSEMBLY—B8784

Figure 18. Barrel Extension Details

The function of the barrel is to direct the discharged projectile. The rifling or grooving causes the projectile to rotate and maintain direction and prevent tumbling.

The barrel is of one piece, threaded at the rear or breech end to screw into the barrel extension. See Figure 17. Although the barrel tapers toward the front or muzzle end, the last portion is ground straight so as to permit it to slide in the front bearing of the barrel jacket. A chamber is formed in the barrel at-the breech end which has the exact contour of the cartridge. A series of notches or serrations is formed in the rear cylindrical outer surface. When the barrel is screwed into the barrel extension, one end of the barrel locking spring (Figure 18) fits into these serrations to prevent any change in the degree of engagement between the barrel and barrel extension during firing. Should adjustment be necessary (described later) it can readily be made against the tension of the barrel locking spring.

## BARREL AND BARREL EXTENSION GROUP

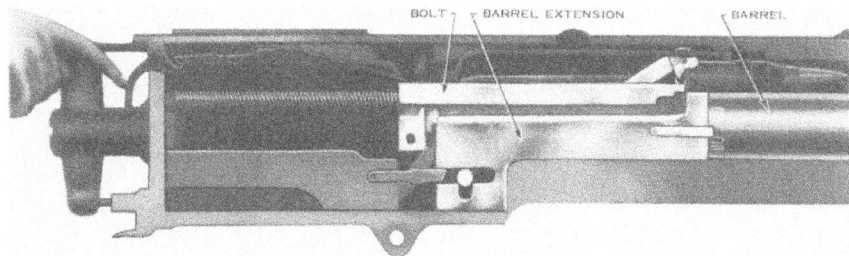

Figure 19 Recoiling Portion

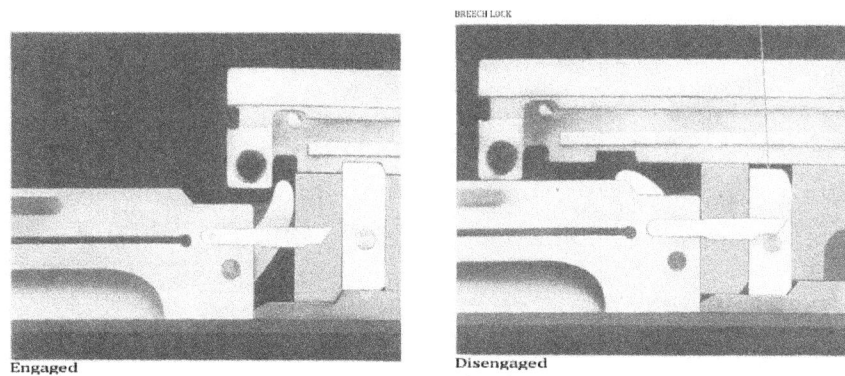

Figure 20 Breech Lock

The barrel extension has lengthwise grooves in which the bolt rides, and further, it houses the breech lock. The breech lock serves to lock the bolt to the barrel extension during and after firing. See Figure 19 and Figure 20.

Fastened to the back end of the barrel extension is the barrel extension shank which engages the oil buffer. The shank is fastened very securely into the extension by a pin.

## OIL BUFFER BODY AND OIL BUFFER GROUP

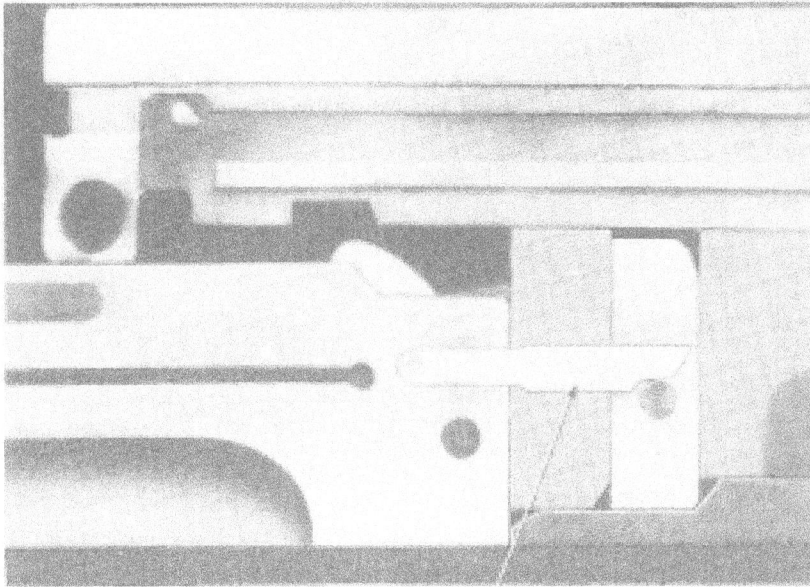

DEPRESSOR

Figure 21 Breech Lock Depressor

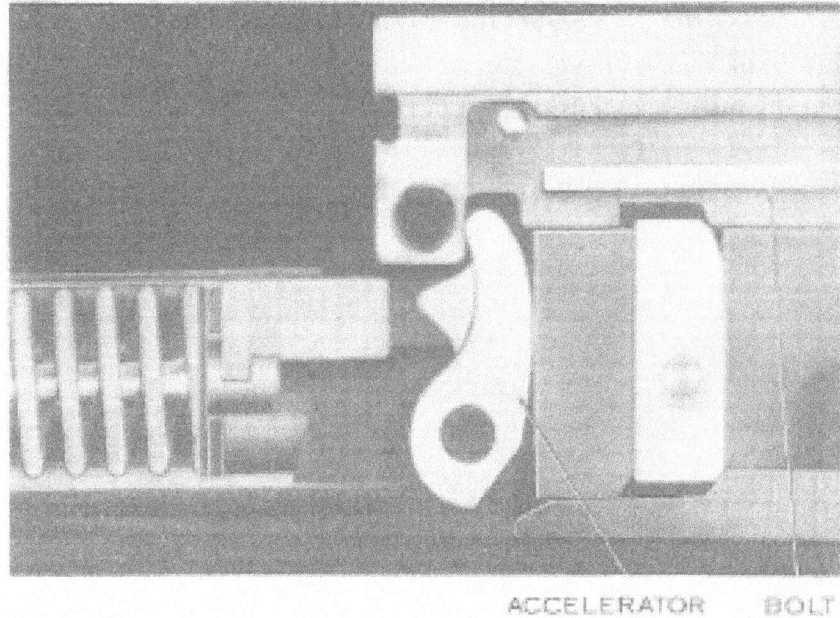

ACCELERATOR    BOLT

Figure 22 Accelerator

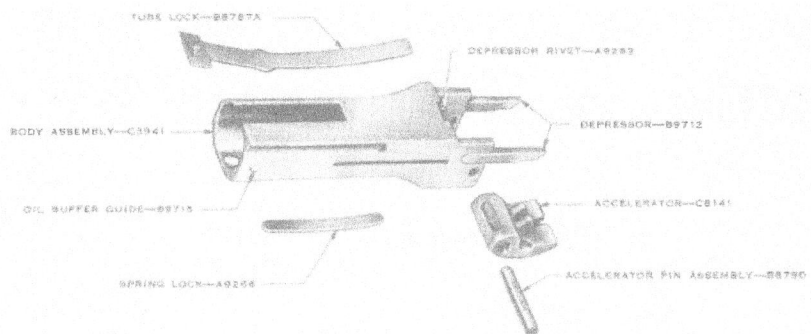

Figure 23 Oil Buffer Body Details

On the recoil or rearward stroke of the barrel extension the breech lock pin is engaged by the breech lock depressors which are riveted to the oil buffer body. See Figure 21. The depressors cause the breech lock to unlock the bolt from the barrel extension.

The accelerator is assembled into the forward portion of the oil buffer body. See Figure 22. On the recoil stroke it assists in driving the bolt to the rear. See Figure 22. During the rearward or recoil stroke the claws on the accelerator bear against the shoulders on the barrel extension shank, thus locking and preventing the barrel extension from moving forward on the counter-recoil stroke until the bolt strikes and moves the accelerator forward. Thus the locking movement of the breech lock is timed so as to bring the lock up exactly when the notch in the bolt is in position.

A flat spring in a groove in the bottom of the oil buffer body (see Figure 23) exerts pressure against the bottom of the accelerator to keep it in the locked position until released by the bolt. The back end of the spring projects beyond the buffer body in such a way that a bevel on the bottom edge of the back plate forces the spring forward. This insures the spring will hold the accelerator very firmly in the locked position; thus the main body of the bolt will pass over the accelerator without interference until the back lug on the bolt actually strikes the accelerator.

## OIL BUFFER BODY AND OIL BUFFER GROUP

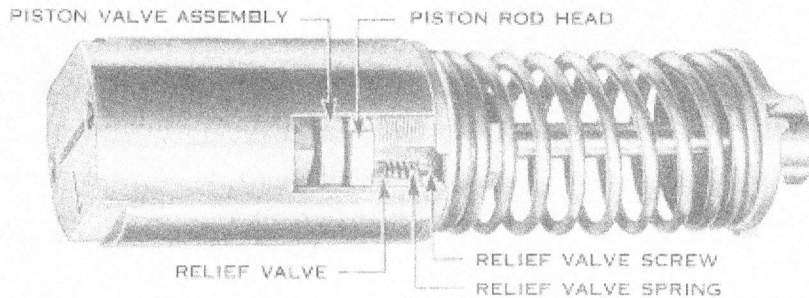

PISTON VALVE ASSEMBLY — — PISTON ROD HEAD

RELIEF VALVE — — RELIEF VALVE SCREW
— RELIEF VALVE SPRING

Figure 24 Oil Buffer Cutaway

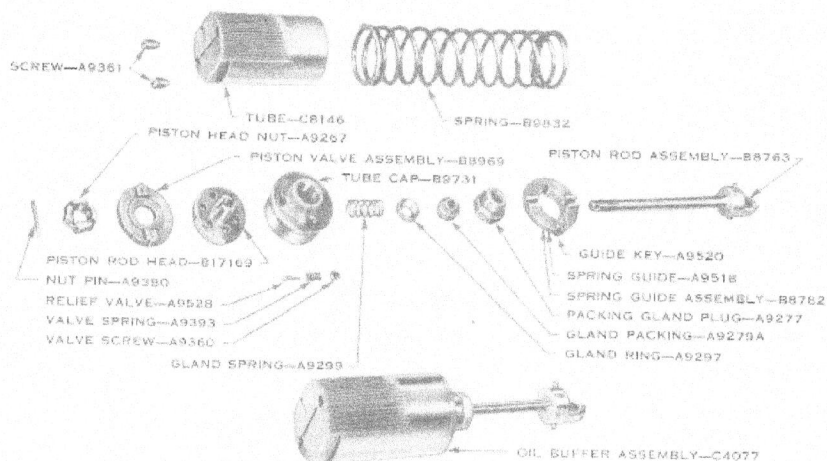

SCREW—A9361

TUBE—C8146
PISTON HEAD NUT—A9267
PISTON VALVE ASSEMBLY—B8969
TUBE CAP—B9731

SPRING—B9832

PISTON ROD ASSEMBLY—B8763

PISTON ROD HEAD—B17169
NUT PIN—A9380
RELIEF VALVE—A9528
VALVE SPRING—A9393
VALVE SCREW—A9360
GLAND SPRING—A9298

GUIDE KEY—A9520
SPRING GUIDE—A9518
SPRING GUIDE ASSEMBLY—B8782
PACKING GLAND PLUG—A9277
GLAND PACKING—A9279A
GLAND RING—A9297

OIL BUFFER ASSEMBLY—C4077

Figure 25 Oil Buffer Details

The oil buffer serves to absorb and partially store the recoil energy of the barrel and barrel extension during the recoil stroke. See Figure 24. This stored energy is given up by the oil buffer spring to drive the barrel extension and barrel forward. The shock absorbing action of the spring is supplemented by a piston and oil cylinder in the rear of the oil buffer spring. See Figure 24 and Figure 25. The degree of oil leakage across the piston on the recoil stroke controls the rate of fire. This may be adjusted manually.

The piston rod and head may slide but are prevented from rotating by the guide key seating in the slot in the oil buffer body. See Figure 25. The piston valve may be rotated to change the leakage aperture at the edge of the piston by turning the oil buffer tube against the restraining action of the lock spring in the tube serrations.

A relief valve in the oil buffer tube cap permits some oil to escape on the initial recoil stroke as the piston rod crowds into the oil filled cylinder. It also allows for oil expansion due to temperature rise.

## RECEIVER AND BARREL JACKET GROUP

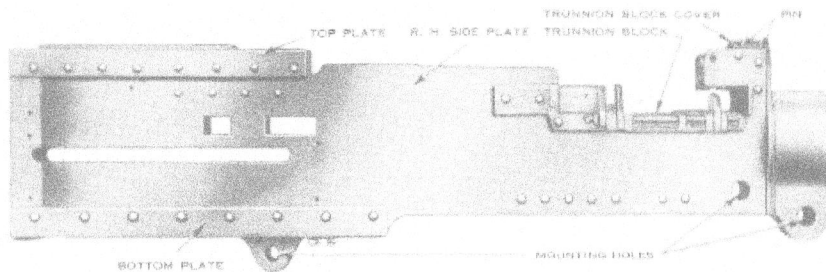

Figure 26. Receiver Assembly

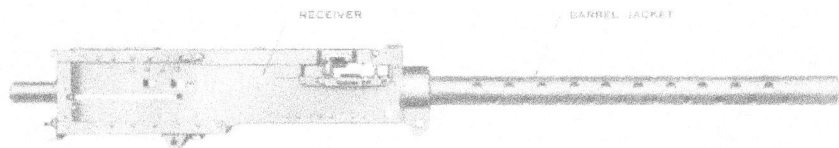

Figure 27 Right Side View of Gun and Receiver

The receiver is probably the most important portion of the gun since it is the "backbone" or main strength member. As such it includes the mountings by which the gun is supported. See Figure 26 and Figure 27. In addition it forms a strong, accurate housing to protect and position the working parts of the gun. It also contains a part of, and supports the remainder of, the ammunition feeding mechanism. It is further utilized to support the various types of trigger mechanisms which are necessary for the different services to which the basic gun may be adapted (i.e., fixed mounting, remote firing as in airplane wing installations, or flexible mounting, manual trigger as in tanks). The barrel jacket is supported by the receiver.

The receiver is made of two steel side plates riveted at their forward portion to a trunnion block; with top and bottom plates riveted to the side plates toward the rear. On top at the extreme forward portion of the receiver a trunnion block cover protects the sight grooves until such time as a sight is installed.

# RECEIVER AND BARREL JACKET GROUP

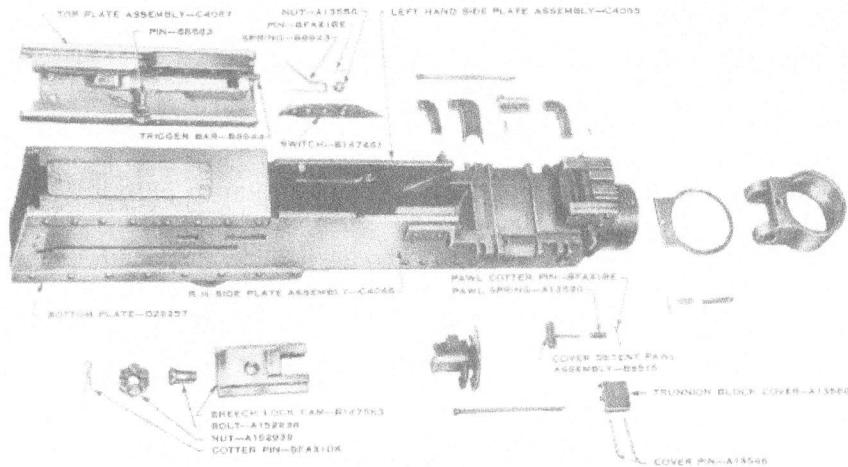

Figure 28 Receiver Details

Directly below the trunnion block cover is a detent pawl which meshes with the cover to retain same in one of three open or raised positions. See Figure 28. The top front of the receiver is open to permit access to the bolt and belt feed mechanism.

Riveted to the underside of the top plate is the top plate bracket which supports the trigger bar pin on which the trigger bar pivots. The trigger bar is assembled in all caliber .50 basic guns even though some applications which demand firing from remote position do not use this piece. Front and rear trigger bar stops are also provided. The top plate bracket has suitable cams for engaging the cocking lever of the bolt. Riveted to the underside of the top plate is the bolt latch bracket which, although part of all basic guns, is used only on those applications which require single shot guns.

The rear of the receiver is slotted to receive the back plate.

The bottom plate carries the breech lock cam which because of a machined shoulder "floats'[7] slightly when bolted down. This cam may have a steel insert or plug in the wearing surface.

The bottom front portion of the receiver is open to permit empty cartridge cases to be ejected.

A switch is pivoted on the inside of the left side plate, with a hairpin spring recessed in a recess in the plate under the switch.

## RECEIVER AND BARREL JACKET GROUP

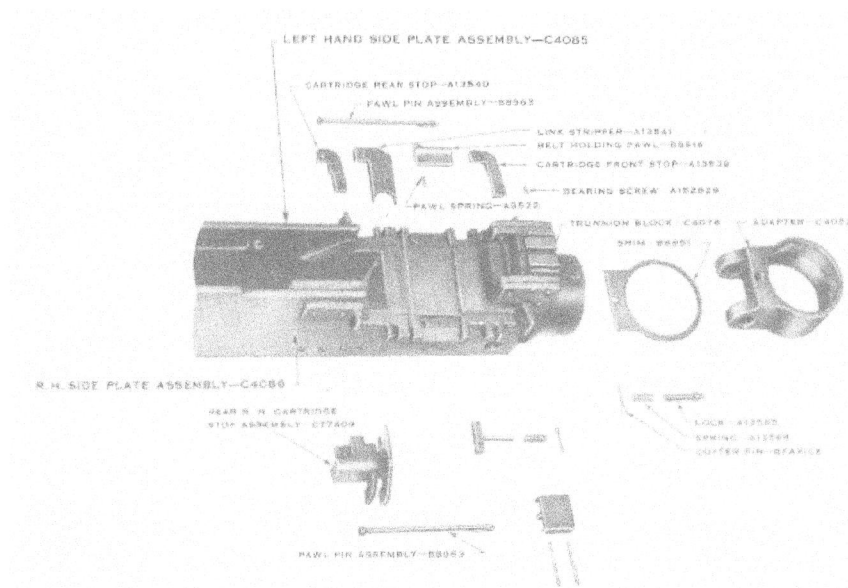

Figure 29 Receiver Details (Front Portion)

The front end of the receiver is formed by the trunnion block which is threaded to fit into the trunnion adapter. See Figure 29. A suitable shim is inserted between the trunnion block and adapter so that the adapter when screwed on tightly will position to line up with the other mountings.

The side plates are notched at the top front portion so that a cartridge belt may be fed into the gun from either side. At these notches the belt holding pawl brackets are riveted to each side plate. These brackets support the belt holding pawl and the cartridge stops, and are so built that parts may be assembled on either right or left side to permit feeding ammunition from either the right or left side. The link stripper and rear cartridge stop are used for right-hand feed only. The rear right-hand cartridge stop assembly is used for left-hand feed only. The cartridge aligning pawl, which is part of this cartridge stop assembly helps to position a cartridge in the feedway when ammunition is being fed from the left-hand side.

## RECEIVER AND BARREL JACKET GROUP

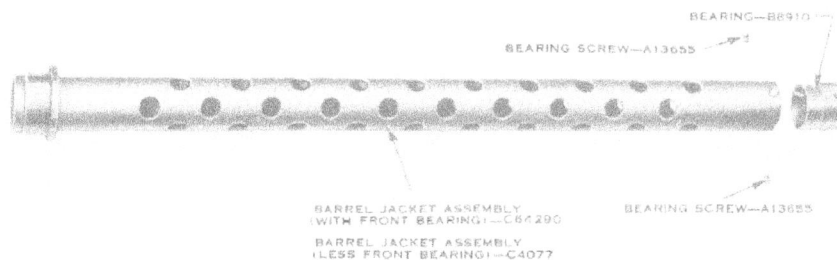

Figure 30 Barrel Jacket Details

The barrel jacket is perforated to permit air to blow through onto the barrel for cooling purposes. See Figure 30. The jacket is stationary, and prevents any object from interfering or rubbing against the barrel which must move during firing. It screws into the trunnion and is locked in place with a small set screw. The front barrel bearing is screwed into the front end of the barrel jacket and is locked in place with two small screws.

## COVER AND BELT FEED GROUP

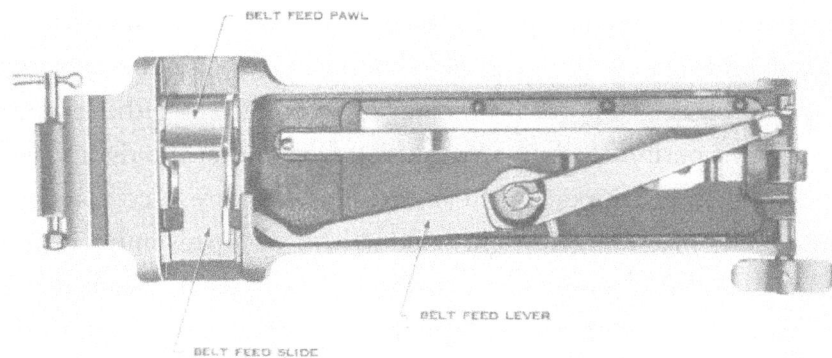

Figure 31

The cover permits access to the bolt, and belt holding parts. On the underside of the cover is the belt feed mechanism. See Figure 31.

The front or hinge end of the cover is serrated so that it may be retained in one of several open positions. A latch is built into the back end of the cover so as to lock it securely to the receiver. This latch may be assembled right or left-hand, depending on the type of slide used.

Operating in a crosswise groove on the underside of the cover is the belt feed slide. This is actuated by the belt feed lever, one end of which rides in the ways on top of the bolt. The belt feed slide carries the belt feed pawl which on each stroke Snaps over a new cartridge and pulls it into position so that it may be extracted from the belt. The pawl, slide, and lever may be reversed to change the direction of feed.

The cover extractor cam is riveted to the underside of the cover. This cam forces the extractor and new cartridge downward as the bolt travels toward the rear. The cover extractor spring which is also assembled to the underside of the cover limits the upward movement of the extractor during the final forward motion of the bolt. See Figure 31A for complete details of cover and belt feed group.

# COVER AND BELT FEED GROUP

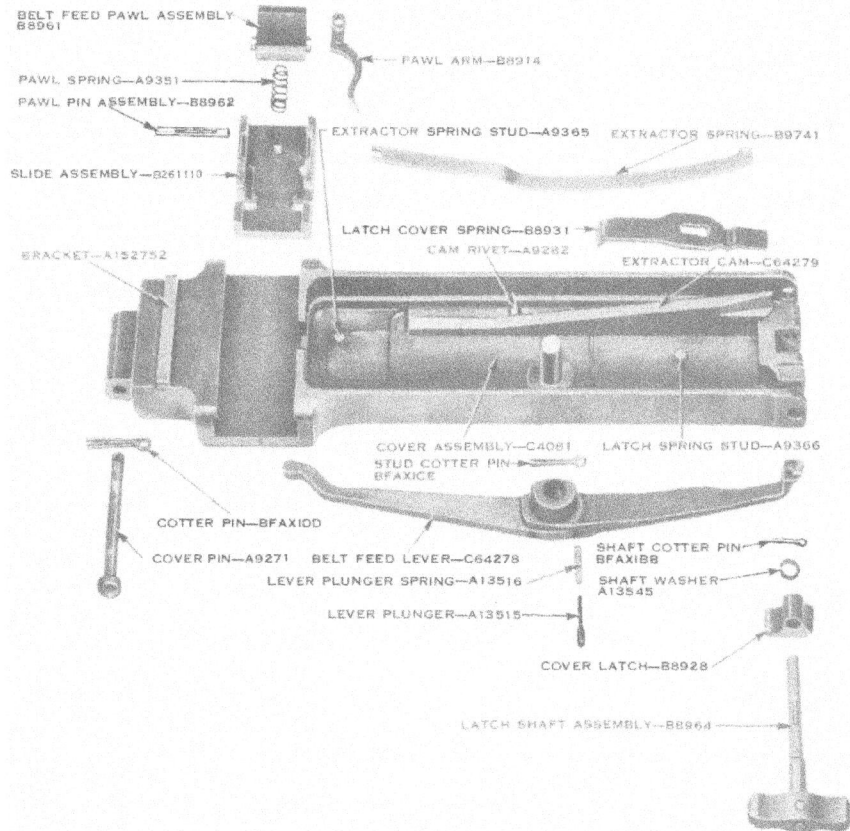

BELT FEED PAWL ASSEMBLY—B8961
PAWL ARM—B8914
PAWL SPRING—A9351
PAWL PIN ASSEMBLY—B8962
EXTRACTOR SPRING STUD—A9365    EXTRACTOR SPRING—B9741
SLIDE ASSEMBLY—B261110
LATCH COVER SPRING—B8931
CAM RIVET—A9282
BRACKET—A152752    EXTRACTOR CAM—C64279
COVER ASSEMBLY—C4081    LATCH SPRING STUD—A9366
STUD COTTER PIN—BFAXICE
COTTER PIN—BFAXIDD
SHAFT COTTER PIN—BFAXIBB
COVER PIN—A9271    BELT FEED LEVER—C64278
LEVER PLUNGER SPRING—A13516    SHAFT WASHER—A13545
LEVER PLUNGER—A13515
COVER LATCH—B8928
LATCH SHAFT ASSEMBLY—B8964

Figure 31A Cover and Belt Feed Details

In the description of the detailed functioning of the caliber .50 Browning Machine Gun which appears on the following pages, it is assumed that, first, the ammunition belt has been properly started into the gun and the cover has been closed and latched, second, the gun has been manually cocked and a cartridge is in its proper position in the chamber and ready to be fired and, third, a manual trigger and trigger bar are to be used to fire the gun.

Each time a cartridge is fired, the mechanical action within the gun involves many parts moving simultaneously or in their proper order. To gain a working knowledge of the operation of these parts and their relationship to each other, the action has been separated into various phases. These are described in the following order:

1. **FIRING**
2. **RECOILING**
3. **COUNTER-RECOILING**
4. **COCKING**
5. **AUTOMATIC FIRING**
6. **FEEDING**
7. **EXTRACTING AND EJECTING**

# FIRING

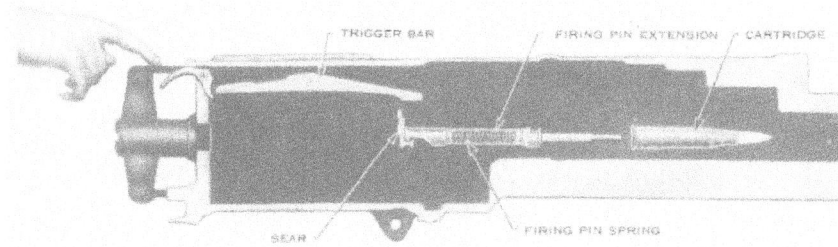

Figure 32

When the gun has been loaded and the firing pin spring has been cocked or compressed manually, the firing mechanism is as shown in Figure 32. The gun is now ready to fire.

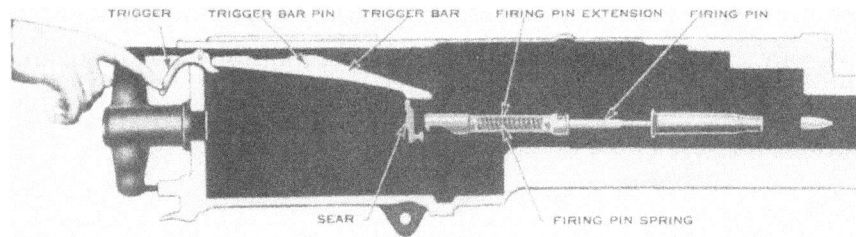

Figure 33

When the trigger is pressed it raises the back end of the trigger bar. The trigger bar pivots on the trigger bar pin, causing the front end to press down on the top of the sear. The sear is forced down until the notch in the sear is disengaged from the shoulder of the firing pin extension. The firing pin and firing pin extension are driven forward by the firing pin spring to fire the cartridge. See Figure 33.

## RECOILING

Figure 34

The complete cycle of the recoiling portion of the gun, which takes place as each cartridge is fired, consists of the recoil stroke when certain parts of the gun move rearward and the counter-recoil stroke when these same parts move forward. At the instant of firing, the barrel, barrel extension, and bolt, known as the recoiling portion, are in the forward position in the gun, as shown in Figure 34.

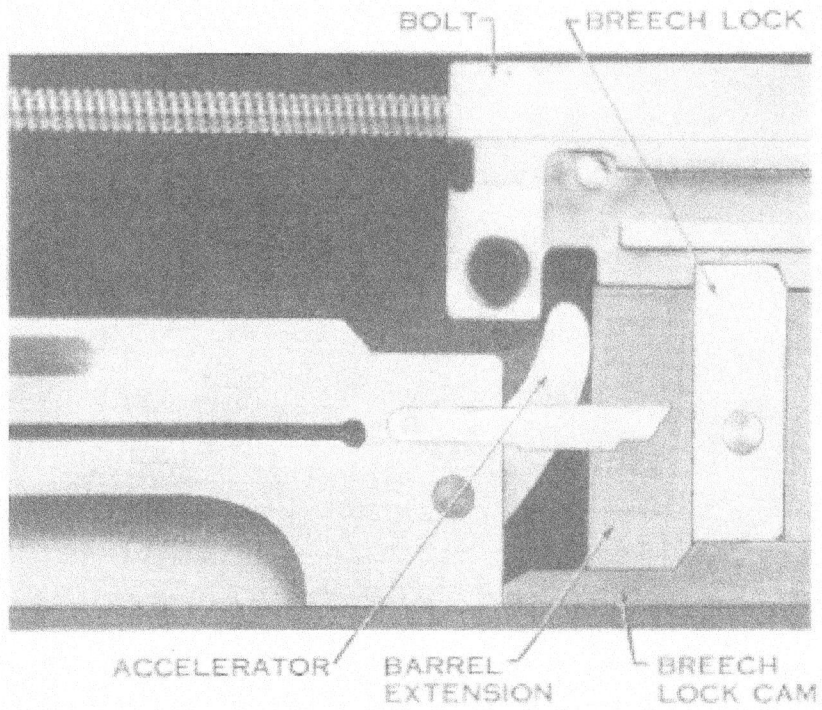

Figure 35

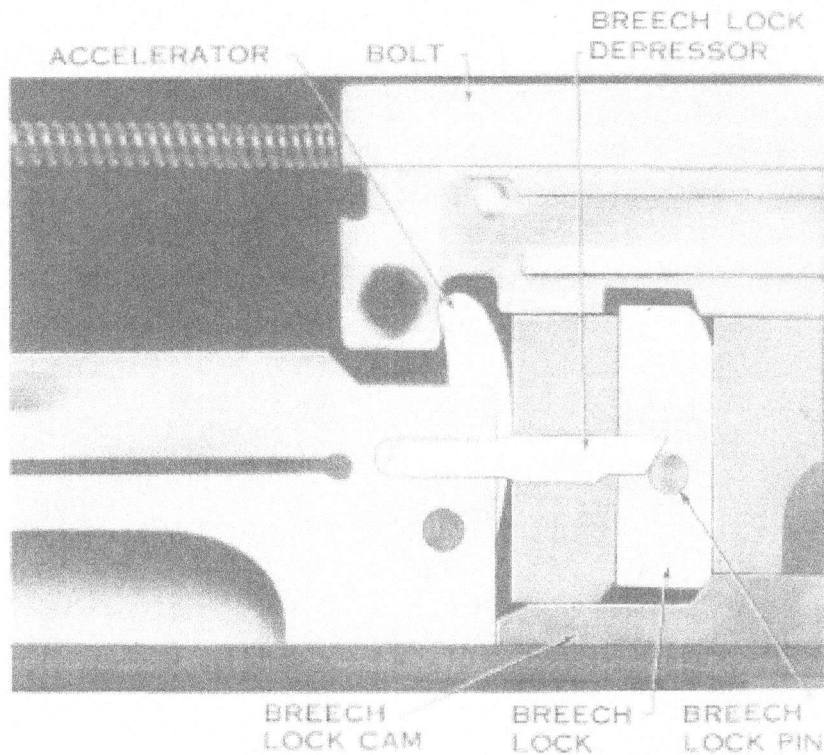

ACCELERATOR　　BOLT　　BREECH LOCK
　　　　　　　　　　　DEPRESSOR

BREECH
LOCK CAM　　BREECH
　　　　　LOCK　　BREECH
　　　　　　　LOCK PIN

Figure 36

At this time the bolt is held securely against the base of the cartridge by the breech lock which extends up from the barrel extension into a notch in the underside of the bolt. See Figure 35.

After the cartridge explodes and as the bullet travels out of the barrel, the force of recoil drives the recoiling portion rearward. During the first three-quarters inch of travel the breech lock is pushed back off the breech lock cam step. See Figure 36. This permits the breech lock to be forced down out of the notch in the bolt by the breech lock depressors engaging the breech lock pin. This unlocks the bolt.

## RECOILING

As the recoiling portion moves toward the rear the barrel extension rolls the accelerator rearward. The tip of the accelerator strikes the lower projection on the bolt and hastens or accelerates the bolt to the rear. See Figure 37. (Note breech lock completely disengaged from bolt notch.)

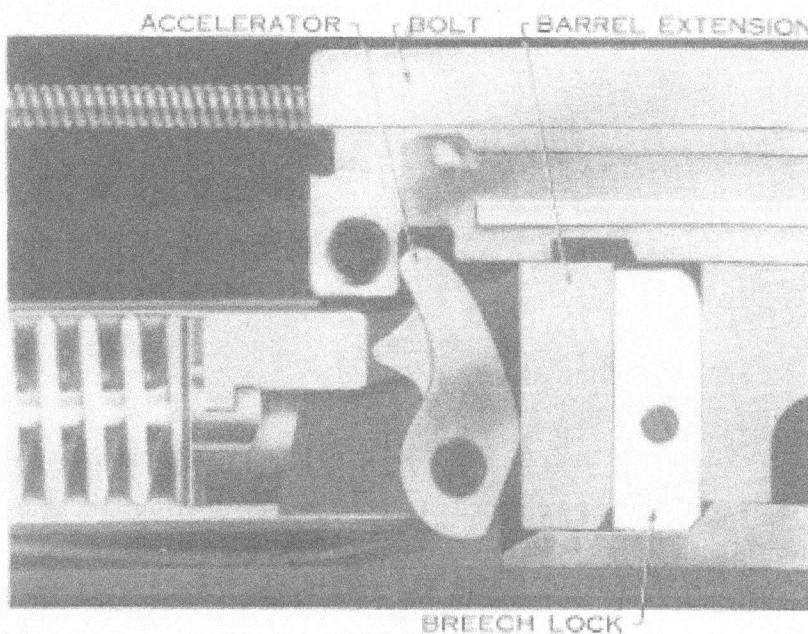

Figure 37

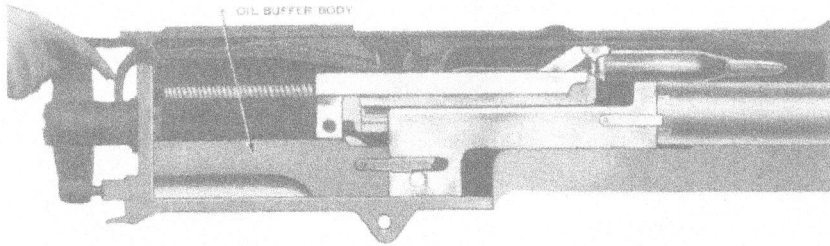

Figure 38

The barrel and barrel extension have a total rearward travel of one and one-eighth inches at which time they are completely stopped by the oil buffer body assembly. See Figure 38.

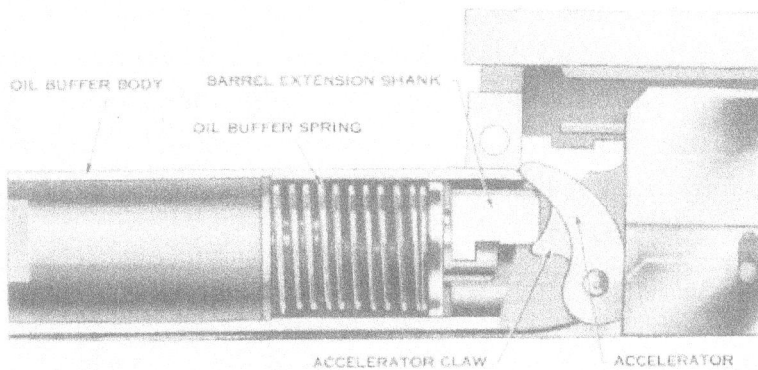

Figure 39

During this recoil of one and one-eighth inches the oil buffer spring is compressed in the oil buffer body by the barrel extension shank. The spring is locked in the compressed position by the claws of the accelerator which are moved against the shoulders of the barrel extension shank. See Figure 39.

# RECOILING

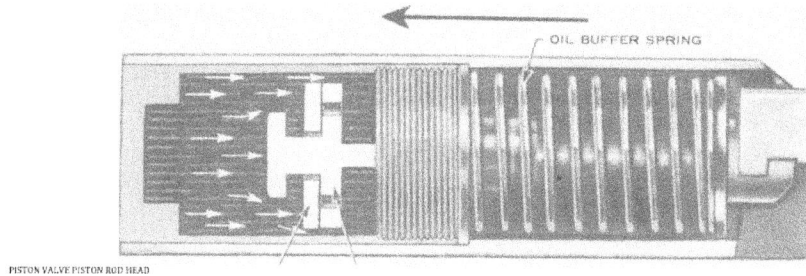

Figure 40

The oil buffer assists the oil buffer spring in bringing the barrel and barrel extension to rest during the recoil stroke, as shown in Figure 40. During the one and one-eighth inch of rearward travel the piston rod head is forced from the forward end of the oil buffer tube to the rear. The oil at the rear of the oil buffer tube under pressure of the piston escapes to the front side of the piston. Its only path is through restricted notches between the edge of the piston rod head and the oil buffer tube.

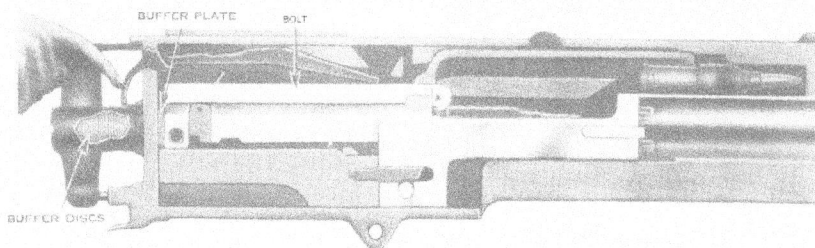

Figure 41

The bolt travels rearward for a total of seven and one-eighth inches. During this travel the two nested driving springs are compressed.

The rearward stroke of the bolt is finally stopped as the bolt strikes the buffer plate, as shown in Figure 41. Thus, part of the recoil energy of the bolt is stored in the driving springs and the remainder is absorbed by the buffer discs in the backplate.

## COUNTER-RECOILING

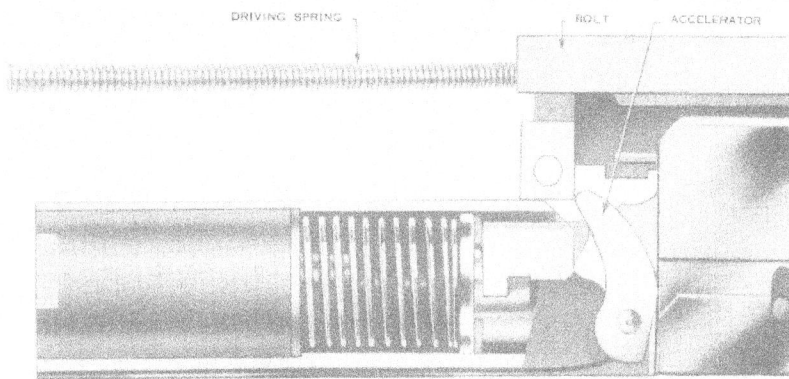

Figure 42

After completion of the recoil stroke the bolt is forced forward by the energy stored in the driving spring and the compressed buffer discs. When the bolt has moved forward about five inches the tip of the accelerator is struck by a projection on the bottom of the bolt. See Figure 42. This rolls the accelerator forward.

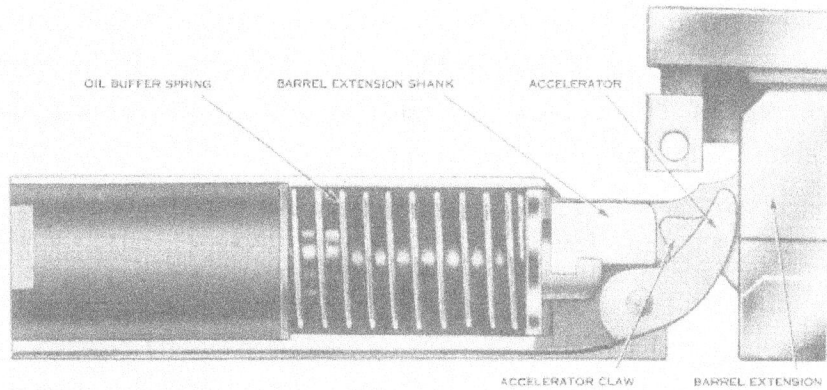

Figure 43

As the accelerator rolls forward the accelerator claws are moved away from the shoulders of the barrel extension shank. This releases the oil buffer spring. The energy stored in the spring shoves the barrel extension and barrel forward. See Figure 43.

## COUNTER-RECOILING

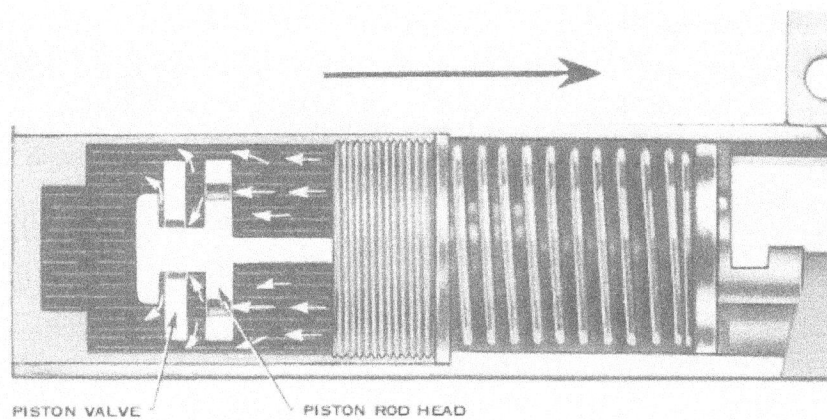

Figure 44

No restriction to motion is desired on the forward or counter recoil stroke of the barrel and barrel extension; therefore, on the forward stroke additional openings for oil flow are provided in the piston rod head of the oil buffer assembly. The piston valve is forced away from the piston rod head as the parts move forward, uncovering these additional openings. This provides an additional path and permits oil to escape freely at the opening in the center of the piston valve as well as at the edge of the piston valve next to the tube wall, as shown in Figure 44.

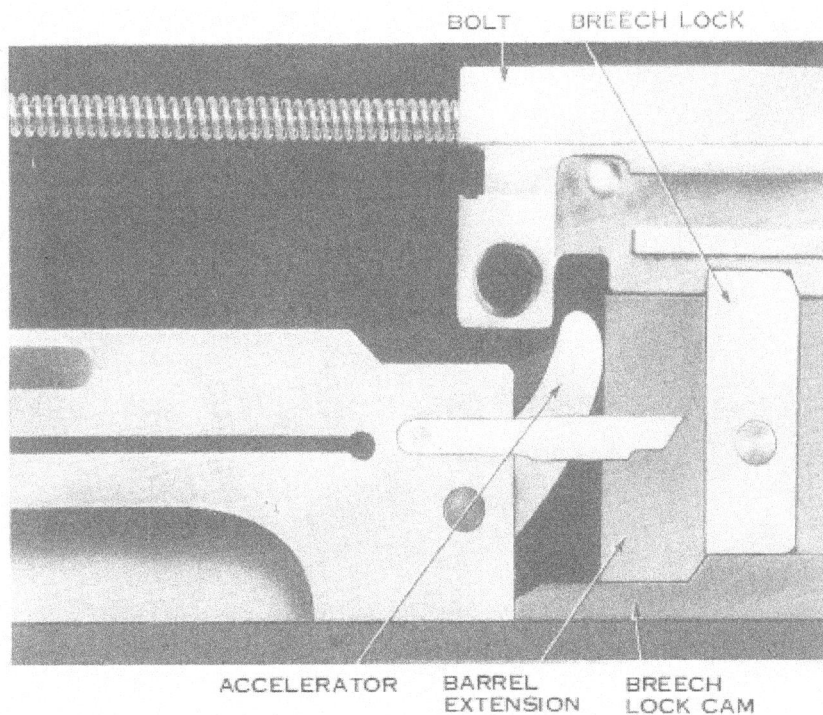

Figure 45

As the barrel extension moves forward the breech lock engages the breech lock cam and is forced upward. The bolt, which has been continuing its forward motion since striking the accelerator, has at this instant reached a position where the notch on the underside is directly above the breech lock, thus permitting the breech lock to engage the bolt. See Figure 45. The bolt is thereby locked to the breech end of the barrel just before the recoiling portion reaches the firing position.

The act of cocking the gun is begun as the bolt starts to recoil immediately after firing. Thus the tip of the cocking lever which is in the V-slot in the top plate bracket, as shown in Figure 46, is forced forward.

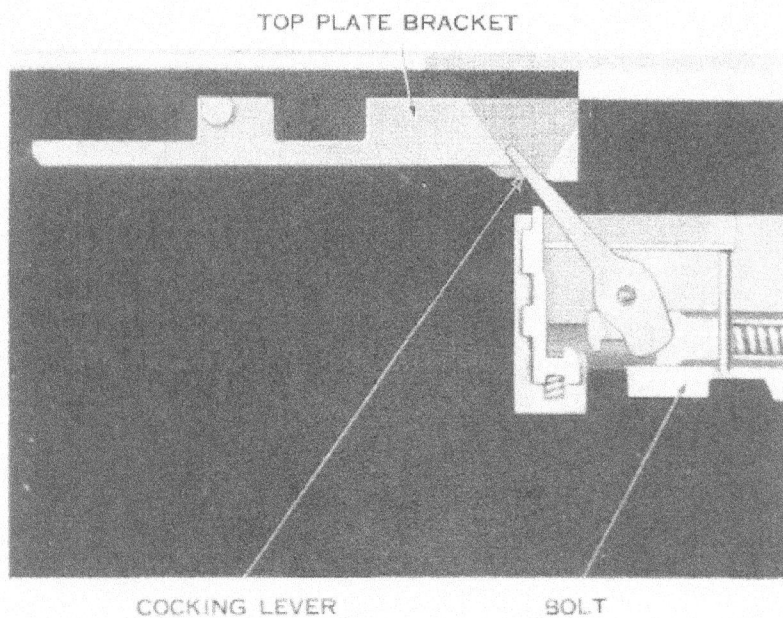

Figure 46

The cocking lever is pivoted so that the lower end forces the firing pin extension rearward. The firing pin spring is thus compressed against the sear stop pin. The shoulder at the back end of the firing pin extension is hooked over the notch at the bottom of the sear under pressure of the sear spring. See Figure 47.

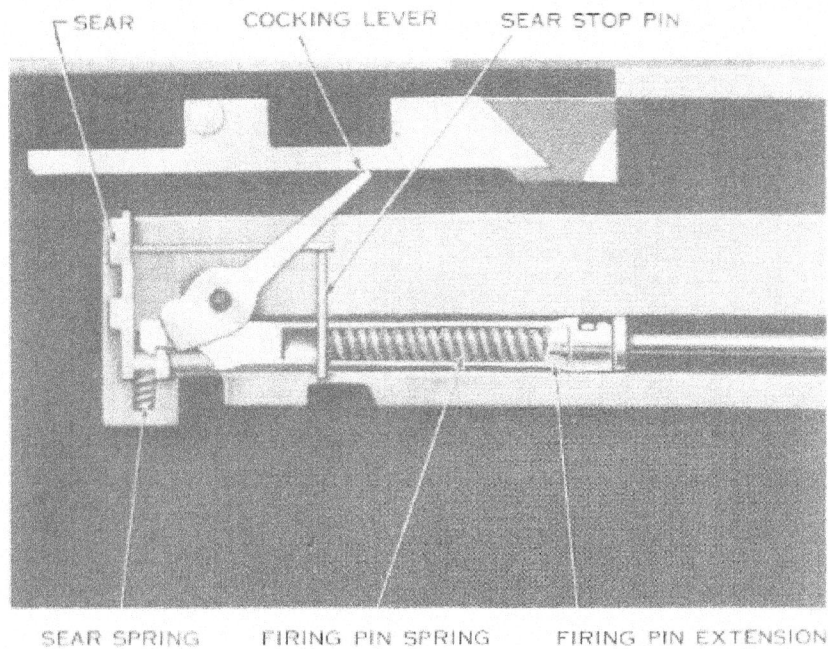

Figure 47

During the forward motion of the bolt the tip of the cocking lever enters the V-slot of the top plate bracket. See Figure 48. This action swings the bottom of the cocking lever out of the path of the firing pin extension, as shown in Figure 49; thus permitting the firing pin to snap forward to fire the cartridge.

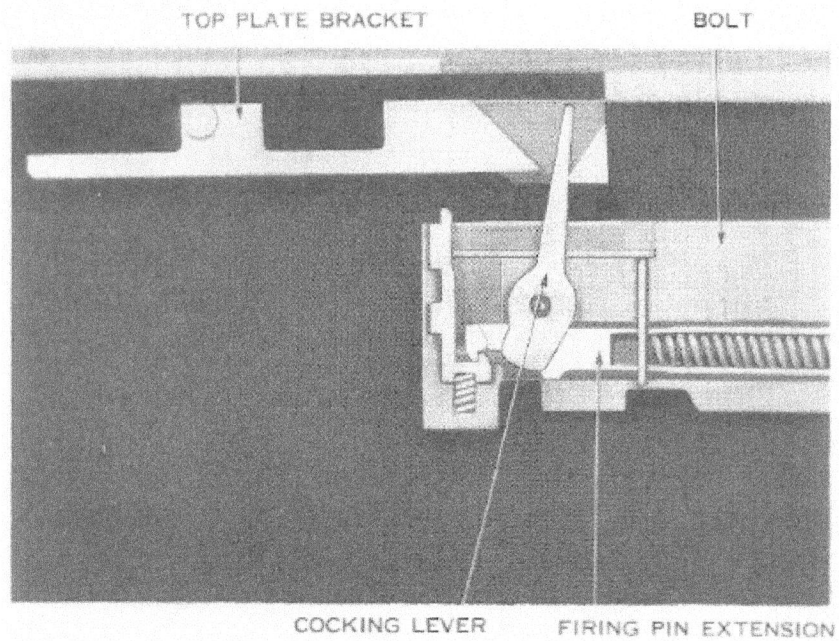

TOP PLATE BRACKET          BOLT

COCKING LEVER      FIRING PIN EXTENSION

Figure 48

When the recoiling portion is almost in the forward position the gun is ready to fire. If no trigger action is given at this instant, the recoiling portion assumes its final forward position, as shown in Figure 49, and the gun ceases to fire. The parts are now in the position shown in Figure 32 and the gun is again ready to fire.

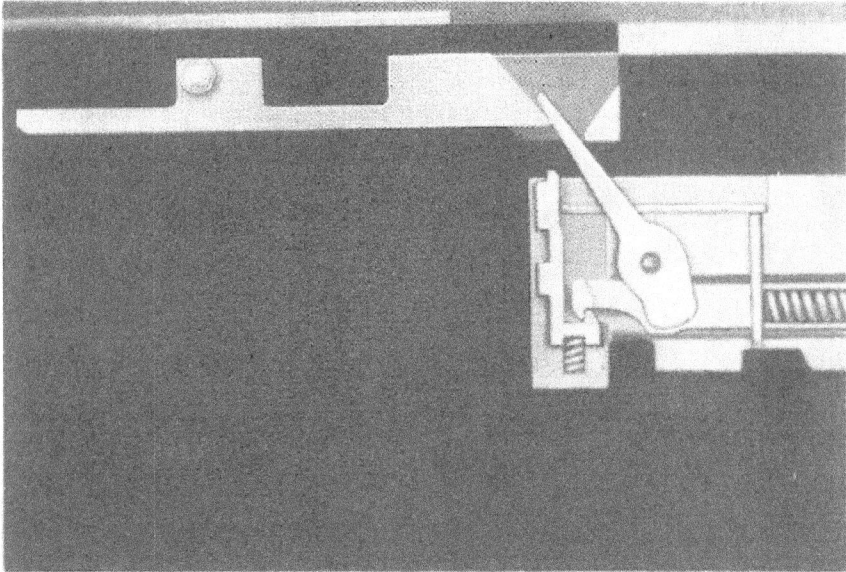

Figure 49

AUTOMATIC FIRING

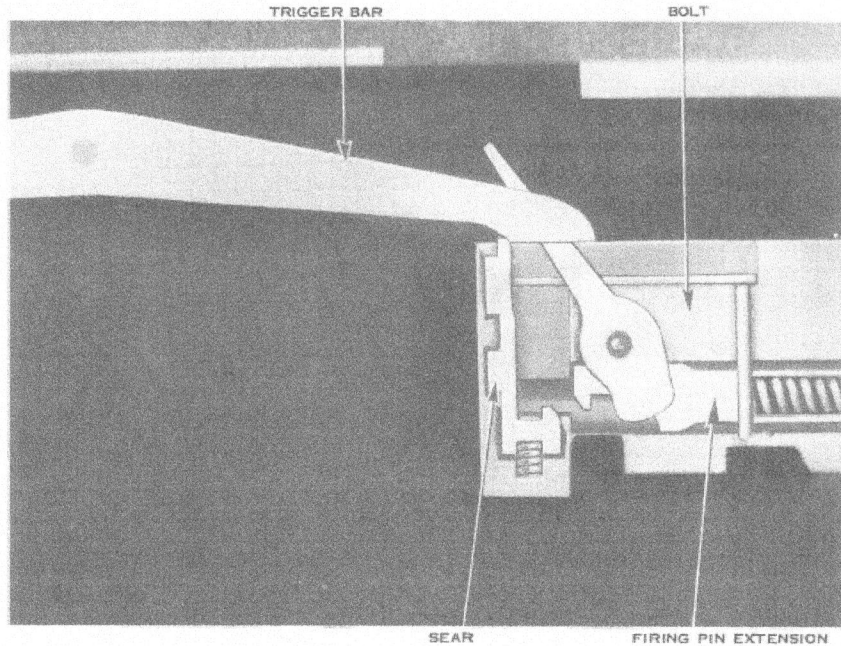

Figure 50

For automatic firing the trigger is pressed and held down. The sear is depressed as its tip is carried against the cam surface of the trigger bar by the forward movement of the bolt near the end of the counter-recoil stroke. See Figure 50. The notch in the bottom of the sear releases the firing pin extension and the firing pin, thus automatically firing the next cartridge at the completion of the forward stroke. The gun fires automatically as long as trigger action is maintained and until the ammunition supply is exhausted.

## FEEDING

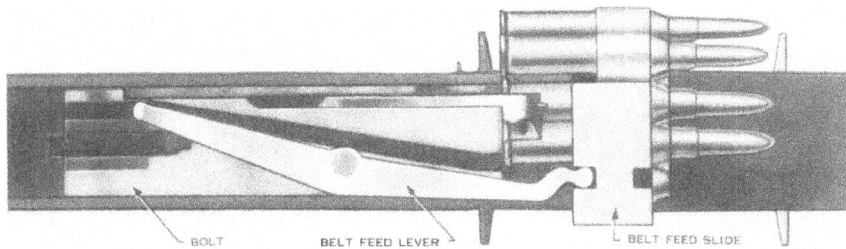

Figure 51

The belt feed mechanism is actuated by the bolt. When the bolt is in the forward position the belt feed slide is within the confines of the gun. Figure 51 shows the mechanism as from above with the cover removed. A stud at the rear of the belt feed lever is engaged in the diagonal groove or way in the top of the bolt.

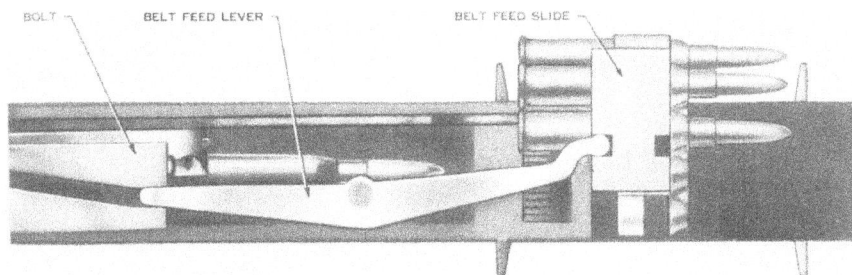

Figure 52

As the bolt moves rearward during recoil the belt feed lever is pivoted. The forward end of the belt feed lever moves the belt feed slide out of the side of the gun and over the ammunition belt. Note: Ammunition feed in Figure 52 is from the left side of the gun. Feed from either side is possible with all caliber .50, M2 guns.

FEEDING

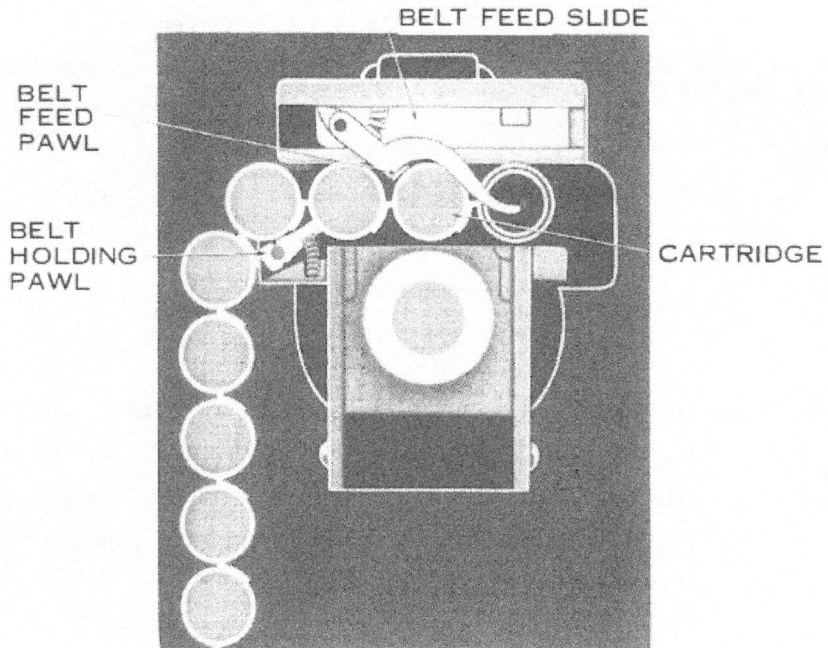

BELT FEED SLIDE

BELT
FEED
PAWL

BELT
HOLDING
PAWL

CARTRIDGE

Figure 53

The ammunition belt is pulled into the gun by the belt feed pawl which is attached to the belt feed slide. When the bolt is forward the belt feed pawl has positioned a cartridge directly above the chamber. The belt holding pawl is in a raised position to prevent the ammunition belt from falling out of the gun. See Figure 53.

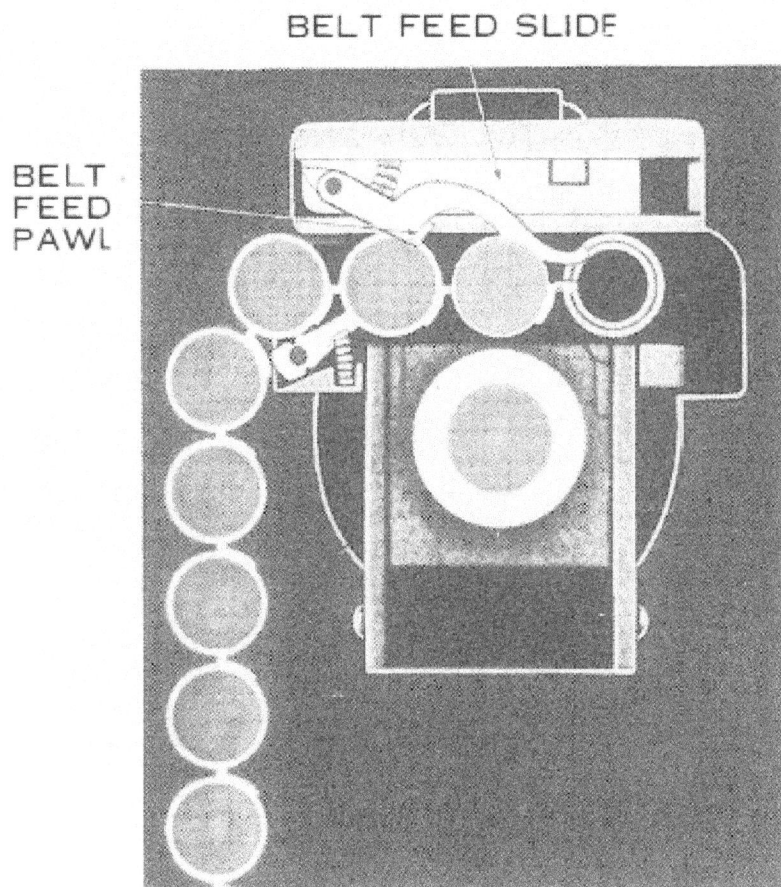

BELT FEED SLIDE

BELT
FEED
PAWL

Figure 54

As the bolt recoils the belt feed slide is moved out over the belt, and the belt feed pawl pivots so as to ride over the next cartridge, as shown in Figure 54.

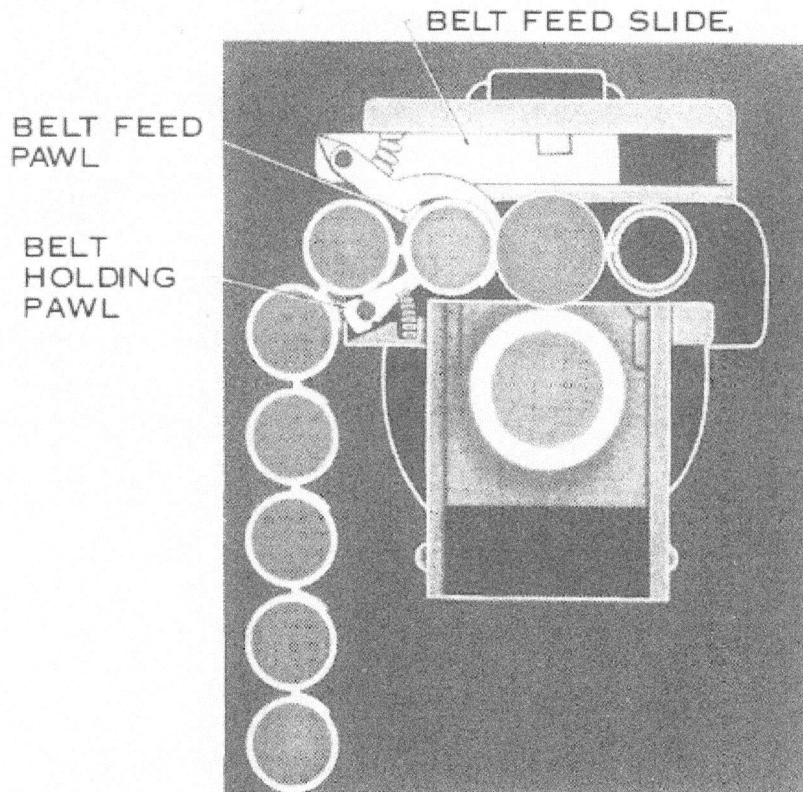

Figure 55

At the end of the recoil stroke the travel of the belt feed slide is suffi-
cient to permit the belt feed pawl to snap down behind the next
cartridge in order to pull the belt into the gun. See Figure 55.

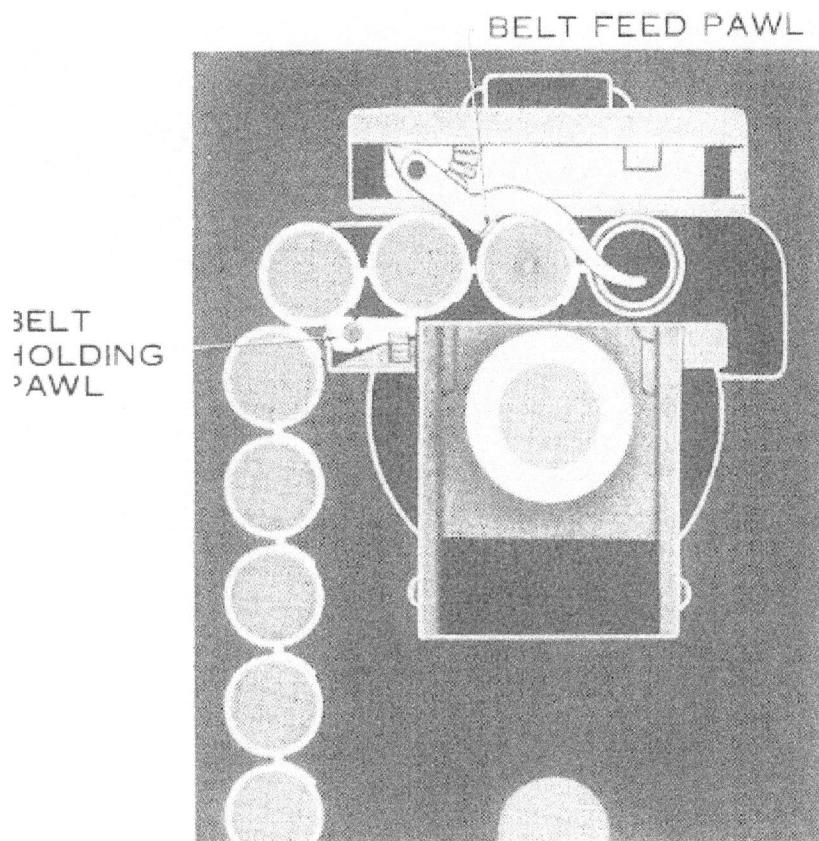

Figure 56

As the bolt moves forward on the counter-recoil stroke the belt is pulled into the gun by the belt feed pawl. The belt holding pawl is forced downward as a cartridge is pulled over it, as shown in Figure 56. When the forward stroke of the bolt is completed the belt holding pawl snaps up behind the cartridge, as shown in Figure 53.

## EXTRACTING AND EJECTING

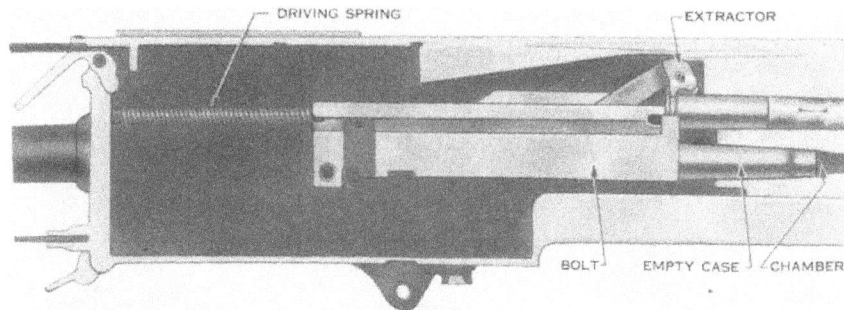

Figure 57

As recoil starts, a cartridge is drawn from the ammunition belt by the extractor. The empty case is withdrawn from the chamber by the T-slot in the front face of the bolt. See Figure 57.

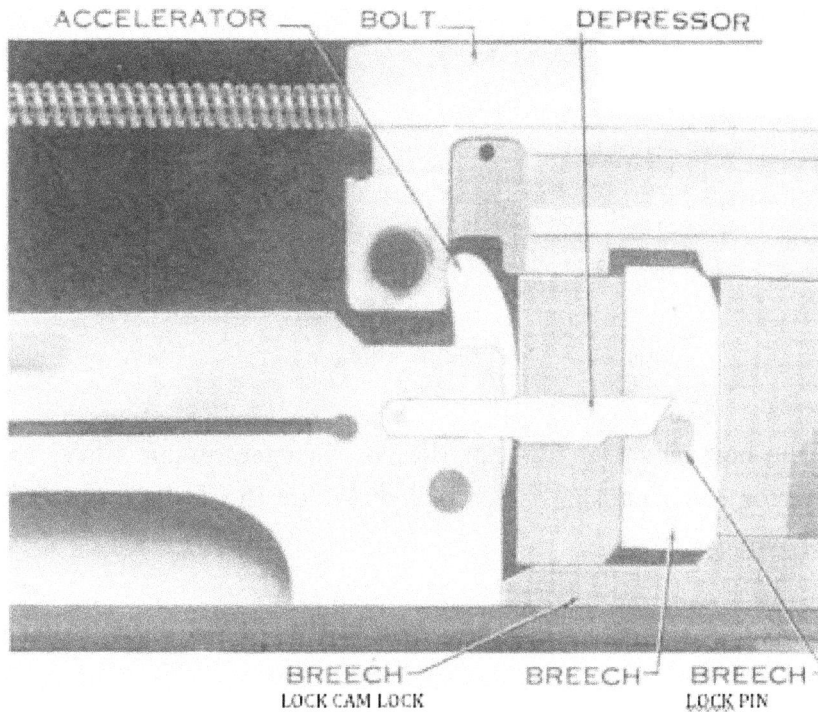

Figure 58

The empty case having been expanded by the force of explosion fits the chamber very snugly and the possibility exists of tearing the case if the withdrawal is too rapid. To prevent this and to insure slow initial withdrawal, the top, front edge of the breech lock and front side of the notch in the bolt are beveled, as shown in Figure 58. Thus, as the breech lock is disengaged, the bolt moves away from the barrel and barrel extension in a gradual manner.

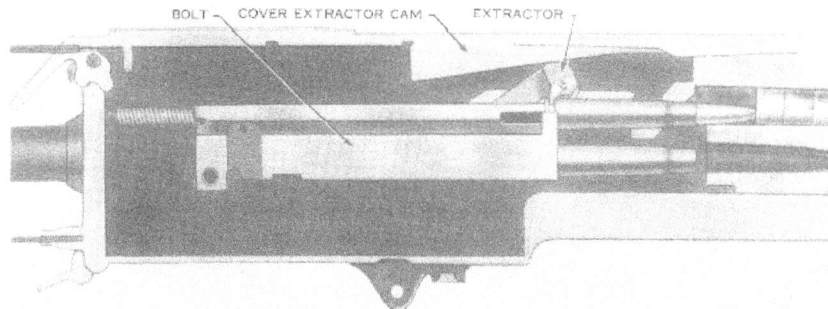

Figure 59

As the bolt moves to the rear the cover extractor cam forces the extractor down, causing the cartridge to enter the T-slot in the bolt, as shown in Figure 59.

## EXTRACTING AND EJECTING

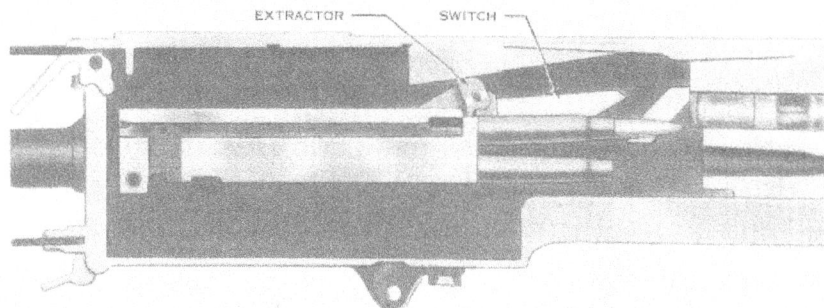

Figure 60

As the extractor is forced down a lug on the side of the extractor rides against the top of the switch causing the switch to pivot downward at the rear, as can be seen in Figure 60. Near the end of the rearward movement of the bolt the lug on the extractor overrides the end of the switch, and the switch snaps up to its normal position.

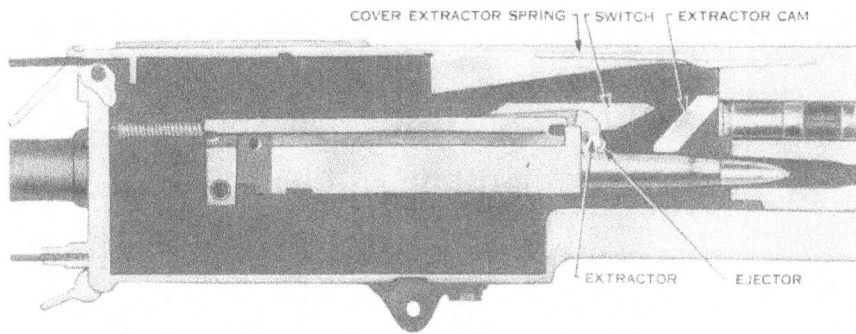

Figure 61

On counter-recoil the extractor and cartridge are forced farther downward by the extractor lug riding on the under side of the switch. The cartridge pushes the empty case out of the T-slot. The extractor stop pin in the bolt limits the downward travel of the extractor so that the cartridge, assisted by the curvature of the ejector, enters the chamber. See Figure 61. (The ejector also ejects the last empty case.) When the cartridge is practically chambered the extractor rides up on the extractor cam, compresses the cover extractor spring, and snaps into the groove in the next cartridge in the belt.

# SAFETY RULES

## (A) SHOP SAFETY RULES:

1. Make sure that the gun is securely anchored in the mountings. Never lay a gun down where it may fall.

2. Make sure that the chamber of the barrel and the T-slot of the bolt are free from cartridges.

3. Never cock the gun against the pressure of the driving spring rod with the back plate removed from the gun.

4. Never use a cloth or waste to apply oil to the working parts of the gun, as lint may be left on the parts which would interfere with the operation. Oil must always be applied with a brush.

5. Never leave twisted ends of locking wires or cotter pins exposed.

6. Never alter or force any part or assembly in such a manner as would tend to make such a part or assembly not interchangeable.

7. Always make sure that the. gun has been checked and adjusted for proper head space.

8. Be sure that the cocking lever always points forward when the bolt is placed in the receiver. See assembly instructions on proper method of assembling the cocking lever in the bolt.

9. Keep tools and bench neat and clean. Maintenance and repair of the caliber .50 machine gun requires careful workmanship, and neatness is the sign of a good workman.

10. After the gun has been function fired, pull the bolt back twice and raise the cover. Check the T-slot and chamber to be sure no cartridges are still in the gun. Release the firing pin spring.

(B) FIRING SAFETY RULES:

1. Before loading the gun, make sure that the bore of the barrel is clear and dry. Chamber and bore must be free of oil.

2. Be sure that the gun has been properly checked for head space.

3. Before loading with live rounds, always test operation by hand using dummy cartridges.

4. Always make sure that the cover is securely latched.

5. Be sure that the back plate is properly in place and that the latch and latch lock are engaged.

6. On a "hang fire" always wait ten seconds before raising the cover to avoid the possibility of the delayed explosion taking place after the cartridge has been removed from the chamber.

7. The adjusting screw of the back plate should be tightened occasionally during firing. Excess play in the discs may result in a broken back plate.

8. After firing, pull the bolt back twice and raise the cover. Check the T-slot and chamber to be sure no cartridges are still in the gun. Release the firing pin spring.

DISASSEMBLY AND ASSEMBLY

## GENERAL DISASSEMBLY:

Disassembly of the caliber .50 machine gun is carried only as far as necessary for instruction, to clean the gun properly, or to make adjustments and repairs. Parts must not be forced into position. To strip the gun of the main groups the cover latch is released and the cover opened. By releasing the back plate latch lock and the back plate latch, the back plate is removed. The driving spring rod is removed by pulling the end to the left, thus releasing the retaining pin from the hole in the side plate. The bolt is drawn to the rear until the bolt stud is in line with the hole in the side plate, "remove stud if a slide is being used" and the bolt is taken out from the rear end of the gun receiver.

Using the point of a cartridge through the hole in the side plate the spring lock is compressed and the oil buffer body, barrel extension and barrel are removed by pulling to the rear. By pressing the accelerator forward the oil buffer assembly is detached. By pressing on the head of the piston rod the oil buffer is pushed out of the body. The oil buffer tube assembly should not be stripped any further than this unless it is necessary to replace the spring because the latter is under sufficient compression to cause serious injury. The barrel may be unscrewed from the barrel extension.

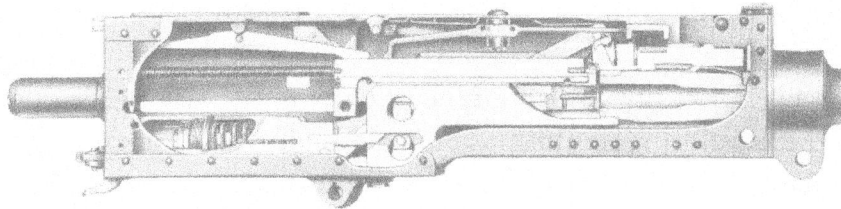

Figure 62. Cutaway View of Gun

## TO DISASSEMBLE THE BACK PLATE GROUP

Before the back plate is removed from its locked position in the gun, the adjusting screw of the buffer assembly may be loosened, using combination wrench No. D-28242, by turning counterclockwise. Care must be taken in removing the screw to avoid losing the plunger spring and plunger.

Release the back plate latch lock and the back plate latch, and remove the back plate from the gun by sliding upward. When the back plate has been removed, the fiber buffer discs and the buffer plate can be readily pushed out by pushing on the buffer plate.

With a drift, force out the filler piece pin and remove the back plate filler piece. The back plate latch is removed by forcing out the back plate latch pin. Again care must be taken to avoid losing the back plate latch spring. Remove the latch lock by taking out the cotter pin and removing the pin. By pushing the sides of the latch lock spring together, the spring can be removed. Detach the lower filler piece by removing the two cotter pins from the filler piece pins, and removing the pins.

## TO ASSEMBLE THE BACK PLATE GROUP

Attach the lower filler piece with the extension to the left, using the two filler piece pins and insert the cotter pins. By pushing the sides of the latch lock spring and inserting the ends in the holes provided in the back plate latch lock with the bowed side toward the latch lock, the spring is properly inserted.

The latch lock is attached to the extension of the lower filler piece by inserting the pin and cotter pin. By inserting the back plate latch spring in the latch, with the other end of the spring in the recess in the lower filler piece, the latch is pressed forward between the thumb and forefinger and the pin is inserted.

The filler piece is placed in the hole and the pin is inserted. Insert the buffer plate in the tube of the back plate with the small diameter entering first. Insert the 22 fiber buffer discs and start the adjusting screw. Replace the back plate assembly in the gun, and, using the combination wrench No. D-28242, tighten the adjusting screw up to the hole in the screw. Then insert the adjusting screw spring and plunger and tighten the adjusting screw securely.

## TO DISASSEMBLE BOLT GROUP (Alternate Feed)

In disassembling the bolt group the extractor is removed by rotating it upward and pulling out frcm the bolt. The bolt switch and bolt switch stud are lifted up. The cocking lever is turned fully backward and by pushing down on the sear the firing pin is released. The cocking lever pin and cocking lever are removed. With the thin end of the cocking lever the sear stop is swung out of its groove. The bolt is turned over, the sear stop pushed out of engagement and removed from its slot. The sear is depressed and the sear slide removed. The sear and sear spring are taken out. The firing pin extension and the firing pin then will slide out to the rear.

## TO ASSEMBLE BOLT GROUP (Alternate Feed)

In assembling the bolt group the firing pin and the extension, with the notch down, are inserted in the bolt and pushed forward until the striker projects through the small hole in the front of the bolt. The sear spring is seated and the sear placed in its guides. In a flexible gun the square end of the sear slide can be either to the right or the left. The slide is inserted and by pressing down on the sear it is engaged. The sear stop is inserted, pushed down as far as it will go, and swung into its recess in the bolt. The cocking lever with the **rounded nose to the rear** is placed in position and held in place by inserting the pin. The correctness of the assembly is tested by pressing forward on the cocking lever, to cock it, returning the lever to its rear position and pressing down on the sear. The click of the firing pin will be heard if the assembly is correct. The bolt switch stud is inserted, the bolt switch is placed over the stud to make the groove marked "L" continuous if left-hand feed is desired, and the extractor inserted in the bolt, being sure the flange is under the collar.

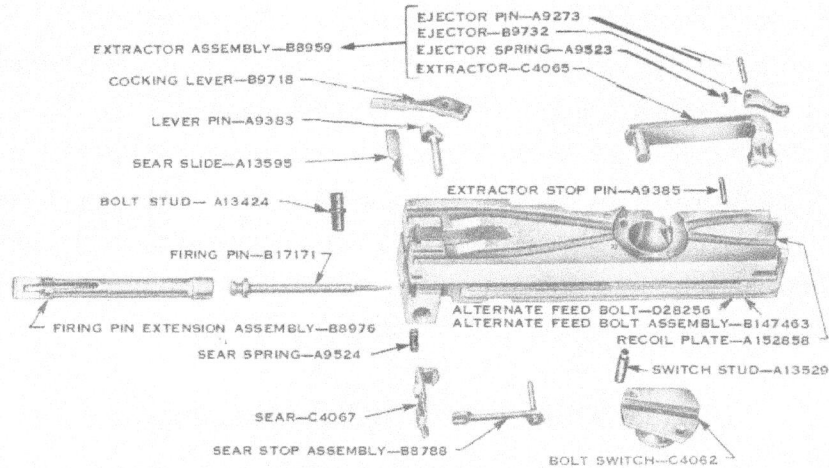

Figure 63 Bolt Group Details

## TO DISASSEMBLE OIL BUFFER AND OIL BUFFER BODY

In disassembling the oil buffer group, the oil buffer body is held bottom up in the left hand. By pressing on the head of the piston rod the oil buffer is pushed out of the body. The index finger is placed between the depressors and the prongs of the accelerator, the stud on the lock spring is disengaged at the same time the accelerator is rotated to the rear: The lock spring then is forced out. The accelerator pin is driven out and the accelerator removed. Remove spring lock from side of buffer body, unless staked in place.

Detach the oil buffer spring by gripping the oil buffer spring guide in a vise, then push on end of the oil buffer until oil buffer piston rod pin clears the projections on the guide. See Figure 64. Carefully give one-quarter of a turn, which will enable pin to pass through the guide. Care should be exercised at this point, as the spring is under considerable compression and may cause an accident if released suddenly. Unscrew oil buffer tube cap and carefully pull out oil buffer piston rod so as not to lose oil. To remove oil buffer piston rod head, take out oil buffer piston head nut pin, then unscrew oil buffer piston head nut, remove oil buffer piston valve and unscrew oil buffer piston rod head.

Unscrew the gland plug from the tube cap. Then oil buffer packing, oil buffer packing gland ring and oil buffer packing gland spring can be taken out.

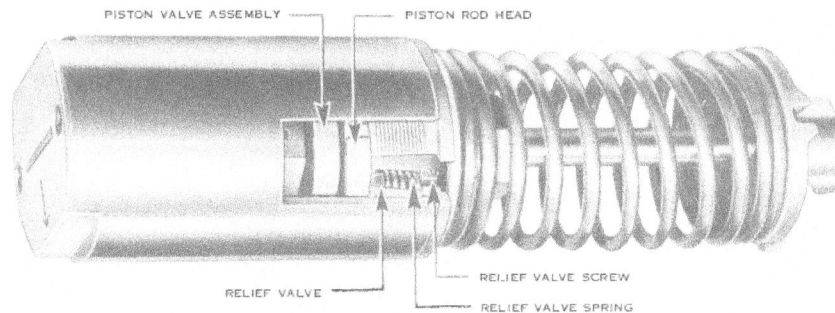

Figure 64. Oil Buffer Cutaway

## TO ASSEMBLE OIL BUFFER AND OIL BUFFER BODY

Place packing gland plug, gland packing, gland ring and gland spring on piston rod in order named, being certain that bevel on packing fits into the gland ring. Screw gland plug into tube cap. Then screw piston rod head onto rod with shoulder away from tube cap. Place valve on rod with flat face toward the piston rod head, and screw the piston head nut on rod until back face of nut is flush with back end of rod. Then screw the piston rod head **back toward the valve** until there is 0.05″ between the faces of these two parts. Insert piston head nut pin through nut, piston head shoulder, and rod, and bend the ends of the pin. Now insert the rod and its assembled parts into the oil buffer tube being careful to line up the valve key with the grooves in the tube wall. See Figure 65.

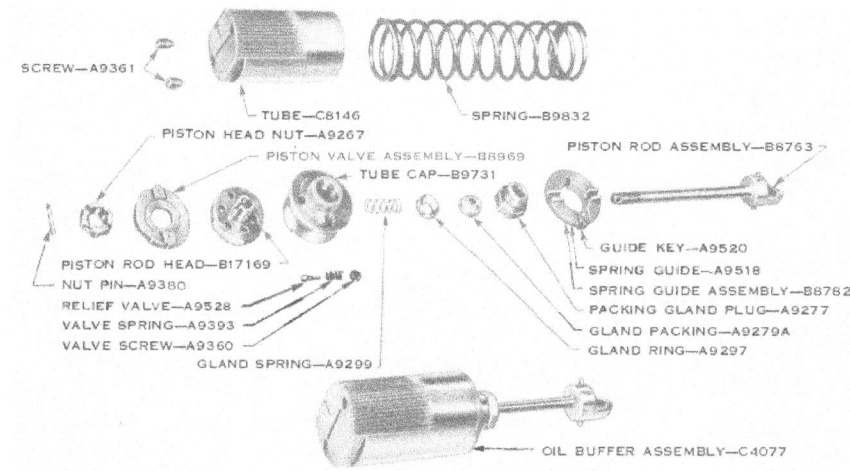

Figure 65. Oil Buffer Details

To assemble the spring to the oil buffer assembly, grip the oil buffer spring guide in a vise with the key up, and place the oil buffer spring over the piston rod. Place the end of the spring against the flat face of the guide and push on the end of the oil buffer until the pin on the end of the rod passes through the slots in the guide. Turn the buffer one-quarter turn and allow the pin to seat in the recesses in the guide. Before removing from the vise, check to see that the flat surface on the piston rod is vertical and the rounded portion is to the right when looking directly at the back end of the oil buffer.

Completely fill the oil buffer with machine gun lubricating oil by removing two filling screws in end of oil buffer; hold buffer upright so that these holes are on top. Start the oil to flow from the spout or nozzle of the oiler and insert the nozzle into one of the holes. Flow oil in until it comes out the remaining hole. Retain pressure on the nozzle while removing it from the hole to avoid possibility of getting air into the buffer. Replace the two filling screws.

Fully insert the oil buffer in the oil buffer body with the guide key engaging in the slot in the buffer body. Replace the spring lock in the side of the buffer body.

In assembling the oil buffer group the accelerator is placed between the depressors with the tips up and rounded surface to the front, then the accelerator pin is inserted, taking care that both ends of the pin are flush with the sides of the body. The lock spring is positioned over the slot in the bottom, depressed into the cut and pushed forward until the stud is seated in the hole.

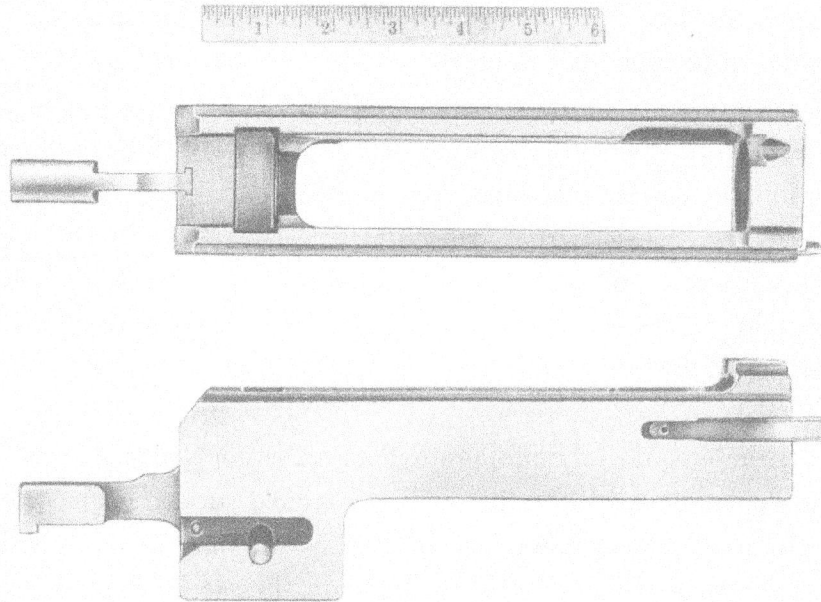

Figure 66. Barrel Extension Group

# TO DISASSEMBLE BARREL AND BARREL EXTENSION GROUP

The barrel is unscrewed from the barrel extension. The locking spring may be removed by sliding it forward out of its seat unless it is staked. See Figure 66. The breech lock pin is pushed out and the breech lock removed.

# TO ASSEMBLE BARREL AND BARREL EXTENSION GROUP

The breech lock is assembled into the barrel extension with the bevel faces to the front and the double bevel on the top. The pin is inserted, taking care that both ends of the pin are flush with the sides of the barrel extension. The locking spring is replaced in the seat and the barrel screwed into the barrel extension. It may be necessary to adjust the degree of engagement between the barrel and barrel extension when the gun is reassembled. See Headspace Adjustment page 55.

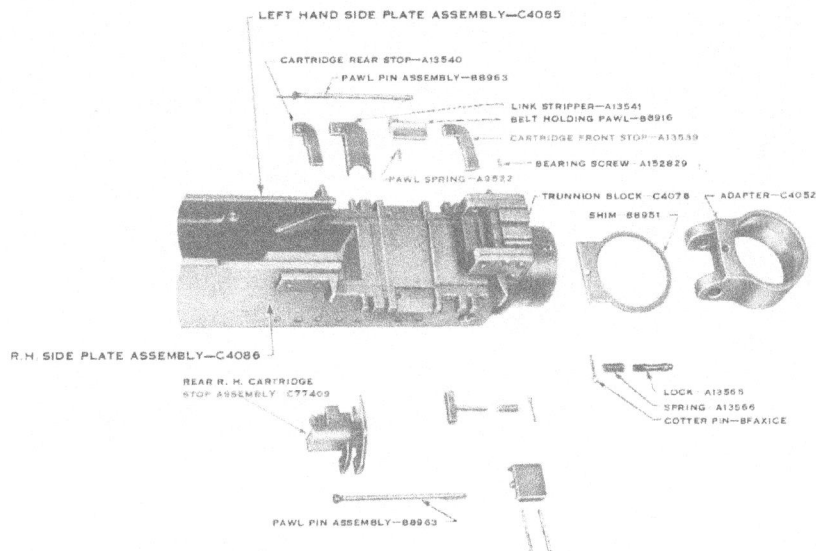

Figure 67. Receiver Details (Front Portion)

## TO DISASSEMBLE THE RECEIVER AND BARREL JACKET GROUP

The belt holding mechanism may be removed from the receiver by withdrawing the belt holding pawl pin, being careful not to lose the belt holding pawl spring. See Figure 67. If the gun is set up for left-hand feed and a rear right cartridge stop is used, it is detached by removing the belt holding pawl pin. It is dismantled by pushing out the pin with a drift. This frees the aligning pawl and permits removal of the aligning pawl plunger and spring. If the rear right-hand cartridge stop is not used, the rear and front cartridge stop and stripper are removed by taking out the belt holding pawl pin.

The cover group is removed by pulling the cotter pin and withdrawing the hinge pin. The cotter pin of the cover detent pawl is pulled to remove the detent pawl and spring.

The trunnion block cover is removed by pushing out two pins with a drift.

The switch and switch spring are removed by pulling the cotter pin and taking off the nut on the outside of the left-hand side plate.

The breech lock cam is removed by taking out cotter pin on the bottom of the receiver and loosening nut. Take out breech lock cam bolt and cam from inside of the receiver.

The top plate cover screws are staked in place so that this usually is not removed. However, the trigger bar is removable. To take out the trigger bar, the trigger bar pin is removed by turning handle downward about one-half inch past the vertical position. Pull the pin out of the side plate.

The trunnion adapter is disassembled, by releasing the trunnion block lock by pulling the cotter pin end of the lock, and at the same time unscrewing the trunnion adapter. This also removes the shim. Withdraw the cotter pin and remove the trunnion block lock and spring.

The barrel jacket is removed by first taking out the breech bearing lock screw from the top of the trunnion, and then unscrewing the jacket with a spanner wrench. The front bearing in the jacket ordinarily is not removed since the two screws are staked.

This leaves the receiver as the only part secured to the bench mounts.

## TO ASSEMBLE THE RECEIVER AND BARREL JACKET GROUP

Assemble the barrel jacket to the trunnion by pulling up firmly with a spanner wrench. If this is a new jacket, a hole for the breech bearing lock screw should then be drilled with a No. 7 drill by mating through the hole in the top of the trunnion. Insert the breech bearing lock screw.

In reassembling the trunnion adapter to the trunnion, insert lock, spring and cotter pin. A shim of one number greater thickness than the one previously removed should be inserted between the trunnion and the adapter. Pull trunnion lock so that lock does not project, and screw on trunnion adapter. If, after drawing up tightly, the cuts in the underside of the trunnion and trunnion adapter do not align, remove adapter and shim and replace shim with next larger number. Repeat if necessary until alignment is secured. The mounting holes in the adapter will then align with the other mounting holes on the gun, and the trunnion block lock will engage itself in the trunnion block adapter.

If possible, turn gun upside down to assist in installing trigger bar. The trigger bar is installed by placing it, hump toward receiver top plate, between the top plate bracket and the bolt latch bracket, with the sear engaging surface forward. The trigger bar pin is inserted, taking care to match the projection on the pin with the notched hole in the left side plate. A bent trigger bar dragging on the face of the bolt will cause sluggish action. This may be checked by drawing several pencil lines across the top of the bolt and observing if these marks are disturbed when the bolt is pulled back.

Push the pin through the hole in the trigger bar and into the hole in the top plate bracket. The pin is turned backwards so as to snap the handle into the small detent recess in the side plate. Install the breech lock cam by laying it in the receiver with the bevel up and rearward. The breech lock cam bolt is inserted downward into the hole in the cam, the nut is drawn up, then backed off about one-sixth of a turn to permit cotter pin insertion.. The cam should "float" slightly.

The switch is installed by first inserting the bent end of the hairpin spring into a small hole in the switch recess in the side plate. The spring is snapped into the recess. The back end of the switch is slid back into the recess, being careful to have the lug on the back of the switch ride on top of the spring. Push the threaded protrusion through the hole in the side plate and secure with nut and cotter pin. Try the switch to see that it pivots and snaps back into position.

The trunnion block cover is set in place and secured with two pins. The detent pawl spring is slipped onto the cover detent pawl assembly and both are inserted into trunnion and secured with a cotter pin. Reinstall cover by placing latch end down against receiver, turning latch, pushing down on cover and releasing latch to catch under the top plate cover. Push hinge end of cover down against detent pawl, and insert hinge pin from right side. Secure with cotter pin.

If the gun is to be equipped for left-hand feed, it is necessary to install a rear right-hand cartridge stop. This stop assembly is prepared by first inserting the spring and then the plunger in the cartridge stop. Then the aligning pawl is hinged in the cartridge stop by inserting the aligning pawl pin. The cartridge stop assembly and front cartridge stop are held to the right belt holding pawl bracket by inserting a belt holding pawl pin assembly.

The belt holding pawl is installed by placing the belt holding pawl spring in the depression in the bracket, holding the pawl in position and inserting a holding pawl pin assembly.

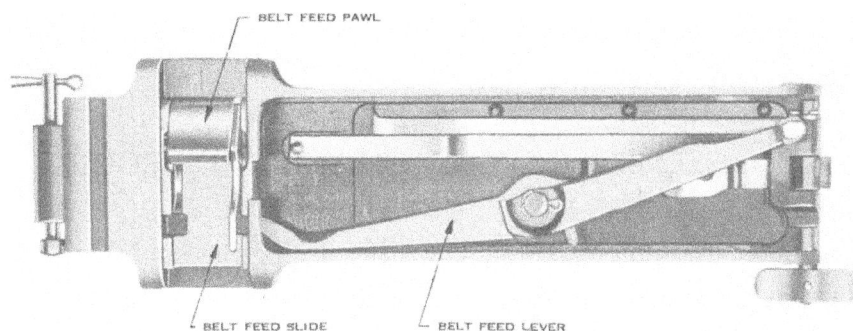

Figure 68. Cover and Belt Feed Group

## TO DISASSEMBLE COVER AND
## BELT FEED GROUP

Open the cover. See Figure 68 and Figure 8. Remove belt feed lever cotter pin and the belt feed lever taking care that the spring and plunger do not fly out. The belt feed lever spring and plunger are then taken out of the hole in the side of the lever.

Remove the belt feed slide from the cover. Push out the belt feed pawl pin and remove belt feed pawl and belt feed pawl arm and spring from the belt feed slide. Keep spring from flying out while doing this.

Lift end of latch cover spring out of the groove in the cover and turn this lifted end slightly so that it rests on the extractor spring. The latch cover spring is then slid away from latch and removed.

The extractor spring is removed by prying the spring away from the cover extractor cam near the latch end.

The latch is taken out by removing the shaft cotter pin and washer, turning the latch shaft to the latched position and withdrawing the shaft from the cover.

## TO ASSEMBLE COVER AND BELT FEED GROUP

The latch is assembled to the cover by placing the latch between the pin bosses on the under side of the cover with the keyway toward the top of the cover and with the projecting wing of the latch against the underside of the cover. The latch shaft is inserted (from either side— depends on type of slide to be used with gun) with the key on the shaft toward the top of the cover. Place washer on end of shaft and install cotter pin. This should be inserted so that the head is toward the hinged portion of the cover, and the ends must be bent sharply to avoid interference when latching down the cover.

Install the extractor spring by hooking the slotted end under the extractor spring stud, with the curved end away from the cover. While holding the stud end in place, press the curved end until it rests on the cover and then slide projection into the recess in the cover extractor cam.

To install latch cover spring, lay the spring inside the cover with the enlarged hole meshing with the latch spring stud. The bent end of the spring should be against the cover. Slide the spring toward the latch, making certain that the latch end of the spring rides up over the projecting wing on the latch. Snap the bent end of the spring into the groove in the cover.

Assume that the gun is to be assembled for left-hand feed. Place the belt feed pawl arm against the side of the belt feed pawl so that the arm will be toward the rear when in the assembled gun with the cover closed. The belt feed pawl with arm and spring is placed in the belt feed slide and the belt feed pawl pin is pushed in place. The belt feed slide is placed in its way or groove in the cover with the pawl end of the slide toward the side from which the gun is to be fed.

Insert the lever plunger and spring in the upper hole in the belt feed lever (for left-hand feed) and place lever, with the shoulder up, on the lever pivot stud in the cover. In order to do this, have the gap in the belt feed slide in line with the cutout in the cover. Push the belt feed lever completely down so that the toe of the lever can work to and fro in the slot provided in the cover. Replace the belt feed lever cotter pin.

## GENERAL ASSEMBLY

Before assembling the gun, the parts should be thoroughly cleaned, oiled, and inspected for burrs, etc. See page 64.

To assemble the gun after it has been stripped, the barrel is screwed into the barrel extension. The oil buffer is fully inserted in the oil buffer body with the guide key in the slot in the body. The barrel extension is held in the left hand and the oil buffer assembly in the right. The accelerator is held up under the shank with the index finger. The breech lock depressors are started in the guide ways of the barrel extension and the oil buffer body is pushed forward as far as it will go.

This complete unit then is pushed into the receiver until it is locked into position. The cocking lever is pressed forward in the bolt and the bolt inserted in the receiver, taking care not to trip the accelerator forward. The bolt is pushed forward until the hole is lined up with the enlarged opening in the side plate and if a slide is to be used the bolt stud is inserted. Inserting the driving spring rod assembly, the bolt is pushed completely forward and the pin is seated in the side plate. The back plate is replaced, the latch lock released and the back plate locked into position. Being sure the bolt is fully forward, the cover is closed and latched and the trigger pressed to relieve the tension on the spring.

The action of the gun should be checked by pulling the bolt back several times to see that all parts are functioning smoothly. The headspace must be adjusted according to instructions on page 55. Operate the gun with dummy cartridges as a final check on correctness of assembly.

# CHANGE FEEDING

To change the ammunition feed from left to right-hand, change over parts as follows:

1. Feed Mechanism:

Open the cover. Remove belt feed lever cotter pin and the belt feed lever.

Change belt feed lever plunger and spring from the upper hole in the belt feed lever to the lower. Do not replace the belt feed lever but lay on bench or table for the present.

Remove the assembled belt feed slide from the cover. Push out the belt feed pawl pin and remove belt feed pawl and belt feed pawl arm and spring from the belt feed slide. Keep spring from flying out while doing this. Change the belt feed pawl arm over from one side of the belt feed pawl to the other, so that when replaced in the belt feed slide it will be to the rear in the assembled gun with cover closed. Replace the belt feed pawl, arm, and spring in the belt feed slide and replace the belt feed pawl pin.

The belt feed slide is always placed in its way with the pawl end of the slide toward the side from which the gun is to be fed.

Replace the assembled belt feed slide in the cover in correct position to feed right-hand. Note that feed pawl arm is to rear. Now replace the belt feed lever in the cover.

In order to do so, have the gap in the belt feed slide in line with the cutout in the cover. Push belt feed lever completely down so that toe of the lever can work to and fro in the slot provided in the cover. Replace the belt feed lever cotter pin.Changing Feed

2. The Bolt:

Remove complete bolt from the gun. Remove the extractor. Raise the bolt switch so as to be clear of the bolt switch stud and give it a half-turn and it will be seen that the other hole in switch will be in line with the stud. Push bolt switch down into place. The way in bolt should now be adjacent to the mark "R" on top of the bolt.

Replace the extractor. Replace the bolt in the gun.

When necessary, reverse position of the sear slide to suit trigger motor or solenoid location.

3. Certain Fittings in the Receiver:

Pull out the two belt holding pawl pins. Take out the front cartridge stop and change to left- hand side. A rear cartridge stop and a link stripper (furnished as auxiliary parts for each gun) should be placed in their grooves on left-hand side. Remove the rear right-hand cartridge stop assembly and retain for time when left-hand feed is again desired.

Change location of belt holding pawl and belt holding pawl spring from left-hand side to right- hand side. Turn the pawl over when doing this. Replace the two belt holding pawl pins.

The changing over of component parts is now completed in order to enable the gun to be fed with cartridges from the right-hand side. To change from right-hand to left-hand feed, reverse the procedure.

## MANIPULATION

TO LOAD:

Before loading, make sure that the bore of the barrel is clear. The end of the belt with the double metallic links is always fed into the gun. Push the end of the belt of cartridges through the feed opening as far as it will go and release. With the belt in this position the first cartridge will be held firmly by the belt holding pawl.

Cock the gun by pulling the bolt to the rear as far as it will go and releasing it. Repeat this operation and the gun is loaded. Firing will continue as long as trigger action is maintained and there are cartridges in the belt.

TO UNLOAD:

Raise the cover. Lift the belt with cartridges out of the feed opening, then close the cover. Pull the bolt to the rear, ejecting the cartridge which remained in the chamber. Check the chamber and the T-slot in the head of the bolt to be sure all cartridges have been removed.

TO ADJUST HEADSPACE:

The headspace of a machine gun with a cartridge fully seated in the chamber is the distance between the base of the cartridge and the face of the bolt.

Correct headspace is important, because, if it is too small, the forward stroke of the bolt cannot be completed and the gun will not fire, and, if it is too large, improper shot patterns will result; and, further, the cartridge case may rupture in the chamber. Headspace is adjusted by obtaining the proper distance between the forward face of the bolt and the rear of the barrel. Headspace adjustment must be checked before firing.

1. Adjustment of Headspace:

In the past the headspace has been adjusted with the barrel, barrel extension and bolt removed from the gun. However, the best adjustment is obtained with the gun fully assembled.

Before starting to adjust the headspace of the gun, the barrel should be unscrewed eight notches from the "all the way in" position. Retract the recoiling portion slightly to expose the locking notches on the breech end of the barrel. Using a screwdriver to engage the locking notches, screw the barrel into the barrel extension one notch at a time and allow the recoiling portion to go forward slowly. Observe whether the forward portion of the barrel extension moves forward far enough to contact the trunnion block.

(This position is commonly referred to as battery position.) Repeat, tightening the barrel, if necessary, one notch at a time, until the barrel extension will go forward and just touch the trunnion block without being forced. Then unscrew the barrel two notches and the headspace will be properly adjusted. CAUTION: Care must be exercised to avoid roughening the barrel notches during adjustment.

2. Checking Headspace Adjustment With the Gage:

After the headspace has been adjusted, the correctness of the adjustment may be tested by using the combination headspace and timing gage, A196228. The portion to be used for checking headspace is marked HEADSPACE—.200, and the portion for checking timing is marked TIMING—.116. The following procedure should be followed in checking headspace adjustment:

(1) When properly headspaced in the manner prescribed above, the barrel will protrude slightly beyond the inner face of the barrel extension.

(2) Cock the firing pin by retracting the recoiling portion of the gun and allow the action to go forward.

(3) Retract the bolt slightly (not more than 1/16 inch) in order to relieve the driving spring pressure to assure that the locking surfaces of the breech lock and its recess in the bolt are in contact.

(4) Then insert the gage in the T-slot between the face of the bolt and the end of the barrel. If the gun is headspaced too tightly, it will not be possible to insert the gage. If such is the case, the barrel

should be unscrewed, one notch at a time, until the gage will just enter the full depth of the T-slot.

(5) If the gun has been headspaced in the prescribed manner, and if the gage will just slide for its full length between the face of the bolt and the end of the barrel without being pushed downward, the headspace is correct. It must be clearly understood that the headspace gage is a "go" gage which was designed particularly for the purpose of checking guns in installations when tight headspace would cause serious trouble.

(6) However, the gage may be used to determine whether headspace is unnecessarily loose by screwing the barrel into the barrel extension, one notch at a time, until the gage will not enter and then unscrewing the barrel one notch so the gage will enter properly.

CAUTION: Never release the firing pin while the gage is inserted in the T-slot or the striker will be damaged. However, after checking and removing the gage, the firing pin should be released.

## TO CHECK TIMING

The purpose of this check is to insure that the gun is not fired too early or too late by any of the various means employed to fire the weapon. In extreme cases of early timing, the gun will fire two shots and then stop because recoil from the second shot started before the extractor could engage the next cartridge in the belt. The gun must NOT fire earlier than .116 inches out of battery. On the other hand, if the gun fires too late while firing automatically, the barrel extension will strike the trunnion block as the recoiling portion moves forward on the counterrecoil stroke. During automatic firing, the gun must fire before the recoiling parts reach a point .040 inches out of battery. Only when the first cartridge of a burst is being fired should the firing pin be released with the recoiling portion in the battery position.

1. Checking to Insure That the Gun is Not Fired Too Early:

Proceed as follows:

(1) Adjust the headspace of the gun.

(2) Cock the firing pin by fully retracting the recoiling portion.

(3) Raise the cover and retract the bolt slightly (!4 inch).

(4) Insert the .116 gage (timing end of the combination gage, A196228) between the front of the barrel extension and the trunnion block with the curved end of the gage over the barrel.

(5) Allow the barrel extension to close slowly on the gage. (Continued on next page)

(6) With the gage in place, an attempt should be made to release the firing pin by operating the solenoid, side plate trigger, or trigger bar. The firing pin should NOT be released. If the firing pin is released, the solenoid should be adjusted so that release does not take place with the gage inserted. In cases where the trigger bar or side plate trigger is used, such a part must be exchanged for another part until one is found which will not release the firing pin with the gage in place.

2. Checking to Insure That the Gun is Not Fired Too Late:

Proceed as follows:

(1) Remove the .116 gage and insert in its place a .040 inch feeler gage. With the feeler gage in place, an attempt should be made to release the firing pin. The firing pin MUST BE released.

(2) If the firing pin is not released, the solenoid must be adjusted or the side plate trigger or trigger bar must be exchanged for another part until one is found which will release the firing pin with the .040 inch gage in place.

Figure 69 Back Plate Moved Up To Show
Adjustment

## TO CHANGE SPEED OF FIRING

It is possible to change the speed of firing by turning the oil buffer tube which is accessible when the back plate is removed. See Figure 69.

Turning to the right so that the arrow moves toward "C" (meaning closed) lessens the speed and turning toward "O" (meaning open) increases the speed.

Turning the arrow two notches to the right from the "O," or open, position will give the maximum speed of firing consistent with proper functioning.

The oil buffer should not be turned so far that arrow goes beyond the lines coinciding with "C" and "O."

# MALFUNCTIONS

A "malfunction" is an improper or incomplete action of some part of the gun or its ammunition resulting in a cessation of fire called a "stoppage."

When describing or determining a malfunction it is often important to designate the approximate position of the moving parts of the gun at the time the stoppage occurs. There are three classifications of stoppages which are generally used.

FIRST POSITION STOPPAGE—One which occurs when all recoiling parts of the gun are in their extreme forward or battery position.

SECOND POSITION STOPPAGE—One which occurs when the recoiling parts are at any position from just out of battery to the position when the bolt is half-way back on either the recoil or counter-recoil stroke.

THIRD POSITION STOPPAGE—One which occurs when the bolt is in any position from half-way back to all the way back on either the recoil or counter-recoil stroke.

As an indication of the usual position of stoppage caused by each particular malfunction, a number in brackets, such as (1), (2) or (3), is shown at the end of each answer. In some cases where a malfunction may cause a stoppage in more than one position, more than one number will be shown to indicate the most common stoppages.

Q. What causes failure to feed?

A. The stoppage which results from failure to feed may be caused by any of the following:

1. IMPROPER HEADSPACE. (2)

2. The BELT FEED LEVER MAY BE DEFORMED due to the stud jumping out of the slot or way in the top of the bolt; or the belt feed stud may be bent or broken causing the bolt to jam. (2) (3)

3. The CARTRIDGE MAY NOT ALIGN WITH THE EXTRACTOR due to a bent belt feed lever giving only limited movement to the belt feed pawl, or the pawl may be deformed so that it slides over the cartridge, or the belt feed pawl spring may be weak. (1)

4. A BROKEN BARREL LOCKING SPRING allowing headspace to change, broken part jamming mechanism. (2)

5. The EXTRACTOR CAM MAY BE BROKEN OR WORN. (1) (2)

6. The EXTRACTOR MAY RIDE TOO HIGH because of a weak or missing extractor spring in the cover. (1)

7. A WEAK DRIVING SPRING might fail to drive bolt with enough force to actuate feed mechanism. (2)

8. A DEFORMED EXTRACTOR, A BROKEN EXTRACTOR STOP PIN or a BROKEN OR DEFORMED EJECTOR may permit the cartridge to be out of line with the chamber. (3)

9. CARTRIDGES MAY BE TOO SHORT for the extractor to reach the extractor groove or cannelure of the cartridge. (I)

10. LOOSE OR WORN MOUNTS may absorb recoil so that driving spring is not fully compressed. Would appear as Number 7. (2)

Q. What would cause the gun to stop firing with the action completed and a live cartridge in the chamber?

A. The sear has not released the firing pin extension. This may be due to one or more of the following:

1. THE TRIGGER BAR MAY BE BENT so that it fails to operate the sear. (1)

2. THE TRIGGER BAR MAY BE BENT so as to strike the rear stop before it strikes the sear, or the rear stop may be too long. (1)

3. The ENGAGEMENT NOTCH IN THE FIRING PIN EXTENSION may be WORN or BURRED. (1)

4. THE SEAR MAY BE BINDING because of burrs on the sear or in the sear recess. (1)

Q. What would cause insufficient recoil with the result that an empty case would fail to be extracted and stoppage would occur?

A. This could be caused by any of the following in the air-cooled gun:

1. THE OIL BUFFER MAY BE SET in the CLOSED POSITION. (1) (2)

2. GRIT OR BINDING BETWEEN THE BARREL AND THE FRONT BARREL BEARING, and between the barrel and its bearing in the trunnion. (2)

3. SIDE PLATES MAY BE SPRUNG together causing binding against the bolt. (2) (3)

Q. What might cause the gun to not quite close or lock on the forward motion so that operation ceases?

A. This could result from one or more of the following reasons:

1. IMPROPER HEADSPACE. (2)

2. BURRS IN THE BARREL EXTENSION prevent free movement of breech lock. (2)

3. THE BELT FEED LEVER MAY BE BENT, thus destroying the feed timing and preventing the action from closing. (2)

4. THE BOLT MAY DRIVE FORWARD SLUGGISHLY due to the bolt binding in the guides of the barrel extension because of burrs or broken parts, or due to a weak driving spring. (2)

5. THE BELT FEED SLIDE OR ITS SLOT IN THE COVER MAY BE BURRED, thus retarding the speed of the bolt. (2)

6. DIRT RAISES THE BREECH LOCK CAM too high for the lock to ride up on the cam. (2)

7. CARTRIDGE DOES NOT SEAT FULLY IN CHAMBER due to burrs on edge of chamber. (2)

8. BELT FEED PAWL ARM MAY BE DEFECTIVE so that it strikes the rear cartridge stop before the action has been completed. (2)

9. BREECH LOCK CAM SCREW PROJECTS INTO FACE OF CAM to interfere with breech lock. (2)

Q. What happens if an operating slide bar gets bent?

A. It binds on the side plate and stoppage results. (2) (3)

Q. What causes the face of the bolt and the firing pin to become eroded by hot gases?

A. The primers are being pierced by being struck with a firing pin which is too long, or by a pin which is sharp or deformed. An undersized firing pin or an oversized firing pin hole may lead to this condition. (1)

Q. If the bolt jams an empty cartridge case against the lower edge of the barrel or barrel extension, what may have caused this?

A. The cartridge case is sticking in the T-slot due to:

1. EXTRUDED PRIMER. (3)

2. NICKED OR BURRED CARTRIDGE CASE. (3)

3. BROKEN EJECTOR failing to push last cartridge case down through the T-slot. (3)

4. ROUGH OR UNDERSIZED T-SLOT. (3)

5. INSUFFICIENT OR SHORT RECOIL. (3)

Q. If a stoppage occurs, the gun fails to eject, and it is necessary to remove empty case by hand and occasionally the case is ruptured, what is wrong?

A. This may result from one or more of the following:

1. The HEADSPACE may be improperly adjusted. (2)

2. The T-SLOT IN THE BOLT MAY BE BROKEN OR DEFORMED. (3)

3. The CHAMBER MAY BE ROUGH, CORRODED OR SCORED. (2)

4. EXTRUDED PRIMER. (3)

Q. What causes misfire?

A. A light blow by the firing pin. Such a light blow may be the result of:

1. A SHORT FIRING PIN. (1)

2. A DEFORMED OR BROKEN FIRING PIN. (1)

3. FIRING PIN BINDING. (1)

4. HEAVY GREASE such as COSMOLINE around the FIRING PIN SPRING and in the FIRING PIN HOLE in the bolt. (1)

5. EARLY TIMING. (1)

Q. What causes the gun to stop with the bolt forward (stroke completed) and no live cartridge in the chamber?

A. The gun has failed to extract a cartridge from the belt for one of the following reasons:

1. The EXTRACTOR MAY BE BENT, BROKEN OR DEFORMED so that it does not fit the extractor groove of the cartridge. (1)

2. The BELT LINK MAY BE DEFORMED OR TOO SMALL so that extractor cannot pull the cartridge out of the belt. (1)

3. The CARTRIDGES MAY BE TOO SHORT for the extractor to reach the extractor groove. These short rounds sometimes are the

result of the edge of the T-slot of the bolt striking the end of the cartridge and driving the case forward over the bullet. (1)

## THE FOLLOWING MALFUNCTIONS MAY CAUSE STOPPAGES IN ANY OF THE THREE POSITIONS.

Q. What happens if cover is not latched or becomes unlatched during firing?

A. Stoppage occurs, probably with breakage of parts.

Q. What might cause the latch not to engage the cover?

A. This might be due to one or more of the following:

1. The LATCH MAY BE BINDING in the cover or against the top plate or the LATCH SPRING MAY BE WEAK OR BROKEN.

2. The SIDES OF THE COVER MAY BE PINCHED or the COVER MAY BE OUT OF LINE so that it strikes the side plates of the receiver.

3. The cover may not close because the BELT FEED PAWL ARM IS BENT to interfere with the link stripper.

4. The LATCH MAY BE WORN OR BROKEN.

5. The COTTER PIN IS IMPROPERLY INSTALLED in the latch shaft.

Q. What causes a portion of the cartridge case to remain in the chamber?

A. This may be due to the following:

1. The HEADSPACE MAY BE IMPROPERLY ADJUSTED. Readjust, and also check for a broken barrel locking spring.

Q. If the metal in the primer of the empty cartridge is extruded back next to the firing pin indent, what causes this, and should anything be done about it?

A. The firing pin is undersize or the firing pin hole in bolt recoil plate is oversize. This condition should be corrected as it may lead to pierced primer.

Q. What causes uncontrolled fire?

A. This may result from the notch on the sear or the notch on the firing pin extension being worn or deformed. It also may be caused by a weak sear spring.

Q. What is meant by hangfire?

A. A delayed explosion probably caused by a light blow by the firing pin due to weak firing pin spring.

Q. What is meant by partial ignition?

A. Sometimes after being fired the powder in the cartridge does not burn completely. Thus the driving force in back of the bullet is decreased and shooting inaccuracy is the result. In extreme cases the bullet may remain in the barrel, which is very dangerous.

Q. If after releasing the trigger mechanism the gun continued to fire (very dangerous), what would you do?

A. The firing could be stopped by twisting the cartridge belt or by moving the retracting handle rearward and holding the bolt out of battery.

Q. What is meant by misfire?

A. If the primer of a cartridge has been indented by the firing pin but is still unfired, this is called a misfire. Although dangerous while remaining in the gun, it is not critically dangerous until removed.

# CHECKING BEFORE AND AFTER FLIGHT

## POINTS TO BE OBSERVED BEFORE A FLIGHT

The following points must be observed before leaving the ground:

1. Wipe bore and chamber of gun barrel.

2. See that adjusting screws are screwed in tight against buffer discs in backplate.

3. Test functioning of gun by hand, using dummy cartridges.

4. Test functioning of operating slide (fixed gun only) or retracting slide (flexible gun only).

5. Oil carefully.

6. See that sight bases are clamped securely in place (flexible gun only).

7. Make sure that ammunition belt is in good condition and properly loaded.

8. Carefully place belt in ammunition chest and see that metallic belt link chutes are in good condition and in proper alignment.

9. Load gun partially or completely as directed.

## POINTS TO BE OBSERVED AFTER A FLIGHT

The following points must be observed as soon after a flight as practicable:

1. Unload gun completely and remove belt from ammunition chest.

2. Clean bore and all working parts. If this cannot be done at once, oil carefully to prevent rust.

3. Release firing pin spring.

4. The armorer must get a detailed account from the gunner or pilot of the gun's behavior in the air. If stoppages have occurred, their cause must be determined and corrected immediately.

5. At the first opportunity, dismount gun; clean, oil, and inspect all parts; make needed repairs and replacements.

6. On assembling, check operation with dummy cartridges and release firing pin spring after insuring that functioning and adjustments are correct.

MAINTENANCE

## CLEANING

All movable parts of the loading and feeding mechanism should be disassembled and thoroughly cleaned with dry-cleaning SOLVENT. The bore should be cleaned with rifle bore CLEANER. After cleaning, the gun should be given a thin protective coating of lubricating OIL for aircraft instruments and machine guns. If the gun has not been fired, but is in an alert condition, it should be cleaned daily with dry-cleaning SOLVENT, thoroughly dried, and a film of lubricating OIL for aircraft instruments and machine guns, applied. Under no circumstances should the gun be allowed to set without cleaning after it has been fired.

When swabbing the bore, the swabbing should be repeated at each cleaning until a clean flannel patch picks up no foreign matter.

## INSPECTION

The purpose of gun inspection is to determine the condition of a gun and any repairs or adjustments that may be required to keep it in proper operating condition.

During all maintenance operations, the gun should be disassembled and parts thoroughly cleaned and inspected for wear, scoring, oil leaks, cracks, burrs, carbon, pitting, corrosion, and rust.

Any burrs or rough edges should be removed by hand honing.

Parts that are broken or worn, or that cannot be satisfactorily improved by hand honing should be replaced.

All springs should be checked and those not within the specification limits shown on page 75 should be replaced.

The threads on all threaded parts should be checked for burrs or roughness.

All moving parts should be checked to see that they move freely. If any bind is present, the cause of the binding should be determined and corrected.

After the gun has been thoroughly inspected and cleaned, it should be carefully dried and covered with a thin film of oil. If the gun is to be stored or shipped, it should be suitably coated with a rust preventive compound after first cleaning with dry-cleaning SOLVENT.

The following cleaning, preserving and lubricating materials are specified by the Ordnance Department.

1. Lubricants—OIL—lubricating for aircraft instruments and machine guns.

2. Cleaners—CLEANER, rifle bore and SOLVENT, dry cleaning.

3. Rust Preventive Compound—compound, rust preventive, light.

The specifications of the above materials are given in regular U. S. Army Specifications.

# AMMUNITION

## CLASSIFICATION

Based upon use, the principal classifications of the ammunition used in this machine gun are:

1. BALL—For use against personnel and light materiel targets.

2. ARMOR PIERCING—For use against armored vehicles, concrete shelters, and similar bullet resisting targets.

3. TRACER—For observation of fire and incendiary purposes.

4. INCENDIARY—For incendiary purposes.

Another type provided for special purposes is dummy ammunition for training (cartridges are inert).

## IDENTIFICATION

Even though the caliber .50 cartridges are not marked or stamped to indicate the type or model, each may be identified as described below. In general, the only stamping on the cartridge is the manufacturer's initials and year of loading which appear on the base of the cartridge case. On lots manufactured prior to 1940, "Cal 50" is also stamped on the base of the cartridge case. However, the marking on all original packing containers, both boxes and cartons, clearly and fully identifies the ammunition except as to grade. In addition to the marking, colored bands painted on the ammunition boxes, and on carton labels, provide a ready means of identification as to type.

When removed from their original packing containers, cartridges may be identified, except as to ammunition lot number and grade, by physical characteristics as described below. Care should be taken not to confuse these original markings with any subsequent markings made with lithographic marking ink, which is used for an entirely different purpose.

1. BALL—Cartridge, ball, caliber .50, Ml, is the standard ball ammunition for this weapon. Cartridge, ball, caliber .50, M1923, is limited standard. All caliber .50 ammunition have bullets with gilding metal jackets (copper colored).

2. ARMOR PIERCING—All models of caliber .50 armor piercing ammunition may be distinguished by the nose of the bullet which is painted black for a distance of approximately ⅞ inch from the tip.

3. TRACER—Caliber .50 tracer ammunition may be identified by the nose of the bullet which is painted red for a distance of approximately ⅞ inch from the tip.

4. INCENDIARY—Caliber .50 incendiary ammunition may be identified by the nose of the bullet which is painted light blue for a distance of approximately inch from the tip.

5. DUMMY—Caliber .50 dummy cartridge may be identified by a hole in the body of the cartridge case.

## LOT NUMBER

When ammunition is manufactured, an ammunition lot number which becomes an essential part of the marking is assigned in accordance with pertinent specifications. This lot number is marked on all packing containers and on the identification card inclosed in each packing box. It is required for all purposes of record, including grading and use, reports on condition, functioning, and accidents in which the ammunition might be involved. No lot other than that of current grade appropriate for the weapon will be fired.

Since it is impracticable to mark the ammunition lot number on each individual cartridge, every effort should be made to maintain the ammunition lot number of the cartridges once they are removed from their original packing. Cartridges which have been removed from original packing and for which the ammunition lot number has been lost are placed in grade 3. Therefore, when cartridges are removed from their original packings they should be marked so that the ammunition lot number may be preserved.

## IDENTIFICATION CARD

An identification card, approximately 7x15 inches, showing the quantity, type, caliber, model, ammunition lot number, and manufacturer is sealed inside the metal liner on top of the ammunition in each box.

## MARKING

Color bands painted on the sides and ends of the packing boxes further identify the various types of ammunition.

The following color bands are used:

Cartridge, armor piercing: Blue on yellow

Cartridge, ball: Red

Cartridge, ball and tracer, in metallic link belt: Composite band of yellow, red, and green stripes (yellow on left, red in center, green on right)

Cartridge, blank: Blue

Cartridge, dummy: Green

Cartridge, tracer: Green on yellow

Cartridge, incendiary: Red on yellow

Carton labels are similarly marked to show the quantity, type, caliber, model, ammunition lot number, and manufacturer. Color stripes similar to those on the packing boxes are also printed on the labels, except that for blank ammunition the label itself is blue and for dummy ammunition it is green.

The number of hits made upon a target by a certain machine gun or group of machine guns when others are firing upon the same target is sometimes determined by coating the tips of the bullets with lithographic ink. The bullets from each weapon or group of weapons are coated with a distinctive color of ink which, upon striking the target, leaves a smear indicating the source of fire. Cartridges which have been so coated must have the ink removed before return to storage.

## CARE, HANDLING AND PRESERVATION

Ammunition boxes should not be opened until the ammunition is required for use. Ammunition removed from the airtight container, particularly in damp climates, is likely to corrode, thereby causing the ammunition to become unserviceable.

The ammunition should be protected from mud, sand, dirt, and water. If it gets wet or dirty, wipe it off at once. Verdigris or light corrosion should be wiped off. Cartridges should not be polished, however, to make them look better or brighter.

The use of oil or grease on cartridges is prohibited.

Do not fire cartridges with loose bullets or other defects.

Ammunition should not be exposed to the direct rays of the sun for any length of time. This is likely to affect seriously its firing qualities.

Whenever cartridges are taken from cartons and loaded into belts, the latter will be tagged so that the ammunition may be identified as to lot number. Tagging is necessary in order to preserve the grade of the ammunition.

## STORAGE

Whenever practicable, small arms ammunition should be stored under cover. This applies particularly to tracer ammunition which is subject to rapid deterioration if it becomes damp, and may even ignite spontaneously. When necessary to leave small arms ammunition in the open, raise it on dunnage at least 6 inches from the ground and cover it with a double thickness of paulin. Suitable trenches should be dug to prevent water flowing under the pile.

If practicable, tracer ammunition should be stored separately from other ammunition.

If tossed into or placed in a fire, small arms ammunition does not explode violently. There are small individual explosions of each cartridge, the case flying in one direction and the bullet in another. In case of fire, keep personnel not engaged in fighting the fire at least 200 yards from the fire and have them lie on the ground. It is unlikely that the bullets and cases will fly over 200 yards.

## AUTHORIZED ROUNDS

The following ammunition of appropriate grade is authorized for use in the Browning Machine Gun, caliber .50, M2, aircraft:

Cartridge, armor piercing, cal. .50, Ml and M2: Standard
    Cartridge, ball, cal. .50, Ml and M2: Standard

Cartridge, dummy, cal. .50, Ml: Limited Standard
Cartridge, dummy, cal. .50, M2: Standard
Cartridge, tracer, cal. .50, Ml: Standard
Cartridge, incendiary, cal. .50, Ml: Standard

# COMPLETE PARTS LIST

## BACK PLATE ASSEMBLY

| Piece No. | Description | Quantity Per Gun |
|---|---|---|
| *C64311 | Plate Assembly—Back (With Horizontal Buffer) | 1 |
| D35276 | Plate—Back | 1 |
| *A152835 | Disc—Buffer | 22 |
| *B8949 | Latch—Back Plate | 1 |
| *A9275 | Pin—Back Plate Latch | 1 |
| *A9356 | Spring—Back Plate Latch | 1 |
| *B147464 | Lock—Back Plate Latch | 1 |
| A13581 | Pin—Back Plate Latch Lock | 1 |
| *A152875 | Spring—Back Plate Latch Lock | 1 |
| *A152753 | Piece—Back Plate Filler | 1 |
| A9275 | Pin—Back Plate Filler Piece | 1 |
| *A152750 | Piece—Lower Filler | 1 |
| BFAX1BB | Pin—Cotter (Latch Lock and Lower Filler Piece) | 3 |
| *A152854 | Pin—Lower Filler Piece | 2 |
| *A152869 | Plate—Buffer | 1 |
| *A152834 | Screw—Adjusting | 1 |
| *A152839 | Plunger—Adjusting Screw | 1 |
| *A9300 | Spring—Adjusting Screw Plunger | 1 |

*Standard Spare Parts and Authorized Assemblies

# ALTERNATE FEED BOLT AND DRIVING SPRING
# ROD ASSEMBLY

| Piece No. | Description | Quantity Per Gun |
|---|---|---|
| *B147463 | Bolt Assembly—Alternate Feed................................ | 1 |
| D28256 | Bolt—Alternate Feed................................... | 1 |
| A9385 | Pin—Extractor Stop..................... | 1 |
| A152858 | Plate—Recoil................................... | 1 |
| A13529 | Stud—Bolt Switch................................... | 1 |
| *B17171 | Pin—Firing................................... | 1 |
| *B8976 | Extension Assembly—Firing Pin................................ | 1 |
| B8946 | Extension—Firing Pin................................... | 1 |
| *A9382 | Pin—Firing Pin Spring Stop........................... | 1 |
| *A9353 | Spring—Firing Pin................................... | 1 |
| | | |
| *C4067 | Sear................................... | 1 |
| *A9524 | Spring—Sear................................... | 1 |
| *B8788 | Stop Assembly—Sear........................... | 1 |
| A9381 | Pin—Sear Stop................................... | 1 |
| B9723 | Stop—Sear................................... | 1 |
| | | |
| *A13595 | Slide—Sear................................... | 1 |
| *C4062 | Switch—Bolt................................... | 1 |
| | | |
| *B9718 | Lever—Cocking................................... | 1 |
| *A9383 | Pin—Cocking Lever................................... | 1 |
| | | |
| *B8959 | Extractor Assembly................................... | 1 |
| *B9732 | Ejector................................... | 1 |
| *A9273 | Pin—Ejector................................... | 1 |
| *A9523 | Spring—Ejector................................... | 1 |
| C4065 | Extractor................................... | 1 |
| | | |
| *C64305 | Rod, Driving Spring, W/Springs, Assembly................. | 1 |
| *C64304 | Rod Assembly—Driving Spring................................ | 1 |
| A152901 | Head—Driving Spring Rod................................ | 1 |
| *A152899 | Pin—Driving Spring Rod Retaining................................ | 1 |
| B147508 | Rod—Driving Spring................................ | 1 |
| *A152900 | Collar—Driving Spring Rod................................ | 1 |
| *A152898 | Pin—Driving Spring Rod Collar Stop................. | 1 |
| *B147510 | Spring—Driving (Inner)................................ | 1 |
| *B147509 | Spring—Driving (Outer)................................ | 1 |
| *A13424 | Stud—Bolt................................... | 1 |

*Standard Spare Parts and Authorized Assemblies.

# OIL BUFFER BODY AND OIL BUFFER ASSEMBLY

| Piece No. | Description | Quantity Per Gun |
|---|---|---|
| *C3941 | Body Assembly—Oil Buffer | 1 |
| C8063 | Body—Oil Buffer | 1 |
| *B9712 | Depressor—Breech Lock | 2 |
| *A9283 | Rivet—Breech Lock Depressor | 2 |
| B9715 | Guide—Oil Buffer | 2 |
| | | |
| *A9266 | Lock—Oil Buffer Body Spring | 1 |
| *B8787A | Lock—Oil Buffer Tube | 1 |
| | | |
| *C8141 | Accelerator | 1 |
| *B8790 | Pin Assembly—Accelerator | 1 |
| A9276 | Pin—Accelerator | 1 |
| *A9357 | Spring—Accelerator Pin | 1 |
| | | |
| *C4077 | Buffer Assembly—Oil | 1 |
| *C8146 | Tube—Oil Buffer | 1 |
| B9731 | Cap—Oil Buffer Tube | 1 |
| *A9361 | Screw—Oil Buffer Tube Filler | 2 |
| *B17169 | Head—Oil Buffer Piston Rod | 1 |
| *A9267 | Nut—Oil Buffer Piston Head | 1 |
| A9380 | Pin—Oil Buffer Piston Head Nut | 1 |
| *A9279A | Packing—Oil Buffer Gland | 1 |
| *A9277 | Plug—Oil Buffer Packing Gland | 1 |
| *A9297 | Ring—Oil Buffer Packing Gland | 1 |
| A9299 | Spring—Oil Buffer Packing Gland | 1 |
| *B8763 | Rod Assembly—Oil Buffer Piston | 1 |
| B9830 | Rod—Oil Buffer Piston | 1 |
| A9379 | Pin—Oil Buffer Piston Rod | 1 |
| *A9528 | Valve—Oil Buffer Relief | 1 |
| *A9360 | Screw—Oil Buffer Relief Valve | 1 |
| *A9393 | Spring—Oil Buffer Relief Valve | 1 |
| *B8969 | Valve Assembly—Oil Buffer Piston | 1 |
| *A9784 | Key—Oil Buffer Piston Valve | 2 |
| B17175 | Valve—Oil Buffer Piston | 1 |
| | | |
| *B8782 | Guide Assembly—Oil Buffer Spring | 1 |
| A9518 | Guide—Oil Buffer Spring | 1 |
| *A9520 | Key—Oil Buffer Spring Guide | 1 |
| *B9832 | Spring—Oil Buffer | 1 |

*Standard Spare Parts and Authorized Assemblies

# BARREL AND BARREL EXTENSION ASSEMBLY

| Piece No | Description | Quantity Per Gun |
|---|---|---|
| *D35348A | Barrel | 1 |
| *C4082 | Extension Assembly—Barrel | 1 |
| D28254 | Extension—Barrel | 1 |
| *B9728 | Shank—Barrel Extension | 1 |
| A9268 | Pin—Barrel Extension Shank Lock | 1 |
| *B8925 | Lock—Breech | 1 |
| *B8784 | Pin Assembly—Breech Lock | 1 |
| A9274 | Pin—Breech Lock | 1 |
| A9357 | Spring—Breech Lock Pin | 1 |
| *B8908 | Spring—Barrel Locking | 1 |

Standard Spare Parts and Authorized Assemblies

# RECEIVER AND BARREL JACKET ASSEMBLY

| Piece No. | Description | Quantity Per Gun |
|---|---|---|
| D35480 | Receiver Assembly | 1 |
| C4076 | Block Assembly—Trunnion | 1 |
| D28264 | Block—Trunnion | 1 |
| A13572 | Plug—Bunter | 1 |
| *B147583 | Cam Assembly—Breech Lock | 1 |
| C64319 | Cam—Breech Lock | 1 |
| A13533 | Plug—Breech Lock Cam | 1 |
| *A152938 | Bolt—Breech Lock Cam | 1 |
| *A152939 | Nut—Breech Lock Cam Bolt | 1 |
| D28257 | Plate—Bottom | 1 |
| BFAXIDK | Pin—Cotter | 1 |
| C4085 | Plate Assembly—Left Hand Side | 1 |
| C4059 | Bracket—Belt Holding Pawl (L. H.) | 1 |
| A9373 | Cam—Extractor | 1 |
| *A9501 | Rivet—Extractor Cam | 2 |
| D28262 | Plate—Side (L. H.) | 1 |
| A13598 | Rivet—Belt Holding Pawl Bracket (L. H. Long) | 3 |
| A9392 | Stop—Bolt (Alternate Feed) | 1 |
| *B147461 | Switch | 1 |
| *A13556 | Nut—Switch Pivot | 1 |
| *BFAXIBE | Pin—Switch Pivot Cotter | 1 |
| *B8943 | Spring—Switch | 1 |
| C4086 | Plate Assembly—Right-Hand Side | 1 |
| B128730 | Bracket—Belt Holding Pawl (R. H.) | 1 |
| D28261 | Plate—Side (R. H.) | 1 |
| A13598 | Rivet—Belt Holding Pawl Bracket (R. H. Long) | 3 |
| A13698 | Rivet—Belt Holding Pawl Bracket (R. H. Short) | 2 |
| C4087 | Plate Assembly—Top | 1 |
| C4061 | Bracket—Bolt Latch | 1 |
| C4070 | Bracket—Top Plate | 1 |
| A9367 | Stud—Top Plate Bracket | 1 |
| D28263 | Plate—Top | 1 |
| A13522 | Rivet—Bolt Latch Bracket (Short) | 5 |
| A9292 | Rivet—Top Plate Bracket | 2 |
| A9391 | Stop—Trigger Bar (Front) | 1 |
| A9390 | Stop—Trigger Bar (Rear) | 1 |
| A13521 | Rivet—Bolt Latch Bracket (Long) | 4 |
| A9387 | Rivet—Bottom Plate | 16 |
| A9500 | Rivet—Top Plate | 16 |
| A9799 | Rivet—Trunnion Block (Short) | 18 |
| A13558 | Rivet—Trunnion Block (Long) | 4 |

*Standard Spare Parts and Authorized Assemblies

# RECEIVER AND BARREL JACKET ASSEMBLY (Cont.)

| Piece No. | Description | Quantity Per Gun |
|---|---|---|
| C77409 | Stop Assembly—Rear Right-Hand Cartridge | 1 |
| C77408 | Stop—Cartridge (R.H. Rear) | 1 |
| B8975 | Pawl—Cartridge Aligning | 1 |
| A13612 | Pin—Cartridge Aligning Pawl | 1 |
| A13611 | Plunger—Cartridge Aligning Pawl | 1 |
| A13613 | Spring—Cartridge Aligning Pawl | 1 |
| *A13539 | Stop—Cartridge Front | 1 |
| *‡A13540 | Stop—Cartridge Rear | 1 |
| *‡A13541 | Stripper—Link | 1 |
| *B8916 | Pawl—Belt Holding | 1 |
| *B8963 | Pin Assembly—Belt Holding Pawl | 2 |
| A152567 | Head—Belt Holding Pawl Pin | 2 |
| B8917 | Pin—Belt Holding Pawl | 2 |
| *A13497 | Spring—Belt Holding Pawl Pin | 2 |
| *A9522 | Spring—Belt Holding Pawl | 1 |
| *B8944 | Bar—Trigger | 1 |
| *B8683 | Pin Assembly—Trigger Bar | 1 |
| A9819 | Key—Trigger Bar Pin | 1 |
| B9786 | Lock—Trigger Bar Pin | 1 |
| A20708 | Pin—Trigger Bar | 1 |
| *B8515 | Pawl Assembly—Cover Detent | 1 |
| A13001 | Guide—Cover Detent Pawl | 1 |
| B8927 | Pawl—Cover Detent | 1 |
| *BFAXIBE | Pin—Cover Detent Pawl Cotter (No. 48 x ⅜") | 1 |
| *A13520 | Spring—Cover Detent Pawl | 1 |
| *C4052 | Adapter—Trunnion | 1 |
| *A152829 | Screw—Breech Bearing Lock | 1 |
| A13588 | Cover—Trunnion Block | 1 |
| *A13565 | Lock—Trunnion Block | 1 |
| A13546 | Pin—Trunnion Block Cover | 2 |
| *B8951 | Shim—Trunnion Block | 1 |
| *A13566 | Spring—Trunnion Block Lock | 1 |
| BFAXICE | Pin—Trunnion Block Lock Cotter (No. 36 x ¾") | 1 |
| B8939 | Cover—Top Plate | 1 |
| A13608 | Screw—Top Plate Cover | 3 |
| *C64290 | Jacket Assembly—Barrel (With Front Bearing) | 1 |
| *C4047 | Jacket Assembly—Barrel (Less Front Bearing) | 1 |
| *B8910 | Bearing—Barrel Front | 1 |
| *A13655 | Screw—Barrel Bearing | 2 |

*Standard Spare Parts and Authorized Assemblies
‡For Right-Hand Feed

# COVER AND BELT FEED ASSEMBLY

| Piece No. | Description | Quantity Per Gun |
|---|---|---|
| *C4081 | Cover Assembly | 1 |
| D28258 | Cover | 1 |
| A152752 | Bracket | 1 |
| BMCXI | Rivet—Countersunk Head (⅛" x 1⅛") | 2 |
| C64279 | Cam—Cover Extractor | 1 |
| A9282 | Rivet—Cover Extractor Cam | 3 |
| A9396 | Stud—Belt Feed Lever Pivot | 1 |
| A9384 | Pin—Belt Feed Lever Pivot Stud | 1 |
| A9398 | Washer—Belt Feed Lever Pivot Stud | 1 |
| A9365 | Stud—Cover Extractor Spring | 1 |
| A9366 | Stud—Cover Latch Spring | 1 |
| *BFAXICE | Pin—Belt Feed Lever Pivot Stud Cotter | 1 |
| | | |
| *B8964 | Shaft Assembly—Cover Latch | 1 |
| A13586 | Lever—Cover Latch Shaft | 1 |
| A13587 | Pin—Cover Latch Shaft Lever | 1 |
| B8930 | Shaft—Cover Latch | 1 |
| A13544 | Key—Cover Latch Shaft | 1 |
| BFAXIBB | Pin—Cover Latch Shaft Cotter | 1 |
| *A13545 | Washer—Cover Latch Shaft | 1 |
| *B8928 | Latch—Cover | 1 |
| *B8931 | Spring—Cover Latch | 1 |
| *A9271 | Pin—Cover | 1 |
| *BFAXIDD | Pin—Cover Pin Cotter | 1 |
| *B9741 | Spring—Cover Extractor | 1 |
| | | |
| *B8961 | Pawl Assembly—Belt Feed | 1 |
| B8913 | Pawl—Belt Feed | 1 |
| A13517 | Pin—Belt Feed Pawl Arm | 1 |
| A13518 | Pin—Belt Feed Pawl Arm Locating | 2 |
| *A9351 | Spring—Belt Feed Pawl | 1 |
| *B8914 | Arm—Belt Feed Pawl | 1 |
| *B8962 | Pin Assembly—Belt Feed Pawl | 1 |
| A13519 | Pin—Belt Feed Pawl | 1 |
| A9357 | Spring—Belt Feed Pawl Pin | 1 |
| | | |
| *B261110 | Slide Assembly—Belt Feed | 1 |
| A147756 | Slide—Belt Feed | 1 |
| A9363 | Stud—Belt Feed Pawl Spring | 1 |
| | | |
| *C64278 | Lever—Belt Feed | 1 |
| *A13515 | Plunger—Belt Feed Lever | 1 |
| *A13516 | Spring—Belt Feed Lever Plunger | 1 |

*Standard Spare Parts and Authorized Assemblies

## SPRINGS

| Name | Piece No. | Free Outside Diameter (Inches) | Wire Diameter (Inches) | Free Length (Inches) |
|---|---|---|---|---|
| Back Plate Latch Spring | A9356 | 0.240 | 0.042 | 1.125 |
| Adjusting Screw Plunger Spring | A9300 | 0.153 | 0.027 | 0.665 |
| Firing Pin Spring | A9353 | 0.295 | 0.059 | 3.218 |
| Sear Spring | A9524 | 0.241 | 0.033 | 0.560 |
| Ejector Spring | A9523 | 0.143 | 0.031 | 0.340 |
| Driving Spring Rod Spring (Inner) | B147510 | 0.287 | 0.039 | 22.000* |
| Driving Spring Rod Spring (Outer) | B147509 | 0.403 | 0.053 | 22.000* |
| Oil Buffer Packing Gland Spring | A9299 | 0.420 | 0.047 | 0.810 |
| Oil Buffer Relief Valve Spring | A9393 | 0.208 | 0.034 | 0.450 |
| Oil Buffer Spring | B9832 | 1.448 | 0.125 | 5.875 |
| Belt Feed Pawl Spring | A9351 | 0.345 | 0.038 | 1.050 |
| Belt Feed Lever Plunger Spring | A13516 | 0.152 | 0.031 | 0.718 |
| Cartridge Aligning Pawl Spring | A13613 | 0.197 | 0.026 | 0.750 |
| Belt Holding Pawl Spring | A9522 | 0.181 | 0.030 | 0.781 |
| Cover Detent Pawl Spring | A13520 | 0.300 | 0.055 | 0.843 |
| Trunnion Block Lock Spring | A13566 | 0.321 | 0.031 | 0.937 |

*Maximum 23.0 Inches. Minimum 21.0 Inches.

# HEAVY BARREL, M2, GUN

The Browning Machine Gun, Caliber .50, HB, M2 is an air-cooled gun having a much heavier barrel than has the aircraft gun. Its general appearance may be noted in Figure 70.

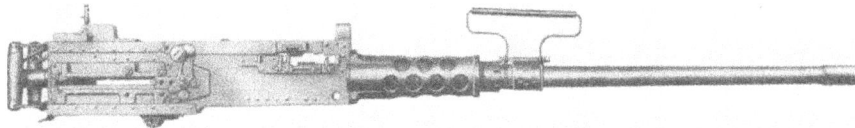

Figure 70. Browning Machine Gun .50, HB, M2

The gun is normally fired in short bursts or in rapid single shots, and when used in this manner firing may be continued for an appreciable length of time because the heavy barrel retards overheating.

## GENERAL DATA (Approximate)

Weight of Gun: 81 lbs.

Weight of Barrel: 27.5 lbs.
Length of Barrel: 45 inches
Number of Lands: 8
Twist—Right Hand: 1 turn in 15
Overall Length of Gun: 65.125 inches
Muzzle Velocity : 2,935 ft. per sec. (2,000 mi. per hr.)
Rate of Fire : 400 to 500 rounds per minute
Maximum Range: 7,200 yards (4.1 miles)

In place of the barrel jacket assembly on the aircraft gun this gun uses a short, perforated barrel support. The trunnion adapter of the aircraft gun is not used. The heavy barrel is removed from the gun by unscrewing it from the barrel extension and withdrawing it toward the front. This permits removing a hot barrel and installing a cool one without disassembling the remaining mechanism of the gun. The handle assembly, shown just ahead of the barrel support, is used for carrying the gun or as a means of turning the barrel when assembling, disassembling or adjusting the headspace. It is moved to one side or down when the gun is being fired. CAUTION: Disengage handle before turning so that headspace adjustment will not be altered.

The firing mechanism is modified somewhat from that included with the aircraft gun. A bolt latch is provided to permit the gun to be fired semi-automatically. It also serves to hold the bolt to the rear in order to keep the cartridge out of the hot chamber when firing has been suspended.

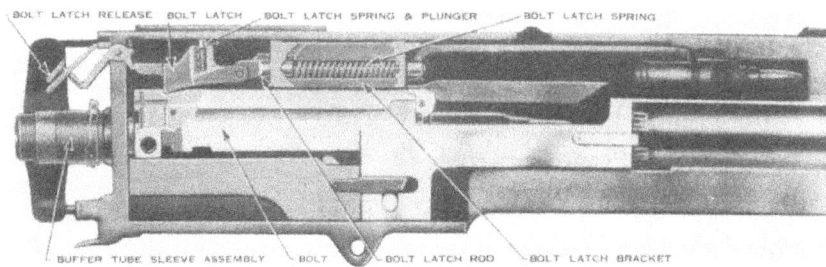

Figure 71. Bolt Latch Mechanism

The bolt latch is forced downward by the bolt latch spring. As the bolt reaches its rearward position, the bolt latch engages a notch on the upper rear surface of the bolt and holds the bolt to the rear, thus causing the gun to cease firing. The counter-recoil stroke is completed by pressing down on the bolt latch release which is pivoted in the back plate. This raises the bolt latch from the bolt notch and allows counter-recoil to take place. Providing a cartridge is in the chamber, firing will be resumed when trigger action is supplied. If the bolt latch release is held down manually, or if it is locked down by the lock on the buffer tube sleeve, the gun will fire automatically. However, if the bolt latch release is pressed down but not retained in that position, the gun will fire only once when trigger action is given.

The back plate spade grip assembly is similar to that used on the Aircraft Flexible Gun except for the addition of the buffer tube sleeve assembly and the bolt latch release and spring.

Since the recoiling portion is much heavier than in the aircraft gun, its rearward motion is not quite so rapid; therefore, it is unnecessary to have as much restriction in the oil buffer on the recoil stroke. Accordingly, the oil buffer piston valve assembly, the gland packing, gland ring, gland spring, oil and oil filler screws are omitted from the heavy barrel gun.

With these exceptions and a few changes in the accessories supplied, such as front and rear sights, the heavy barrel gun is identical with the aircraft gun.

## WATER COOLED, M2, GUN

The Browning Machine Gun, Caliber .50, M2, water-cooled, has a water jacket surrounding the barrel for the purpose of preventing barrel overheating when firing for prolonged periods. The general appearance of this gun may be noted in Figure 72.

Figure 72. Browning Machine Gun, Caliber .50, M2, Water Cooled

## GENERAL DATA (Approximate)

Weight of Gun (with water): 121.5 lbs.
Weight of Gun (without water): 100.5 lbs.
Weight of Barrel: 16 lbs.
Length of Barrel: 45 inches
Number of Lands: 8

Twist—Right Hand: 1 turn in 15 inches
Overall Length of Gun: 65.93 inches
Muzzle Velocity : 2,935 ft. per sec. (2,000 mi. Per hr.)
Rate of Fire : 600 to 750 rounds per minute
Maximum Range: 7,200 yards (4.1 mi.)

The water jacket contains ten quarts of water, and is kept filled by a hand pump from an auxiliary water chest which has a capacity of about eight gallons. During firing, heat absorbed from the barrel changes some of the water to steam. This is removed from the jacket with the water being returned through the jacket outlet to the water chest.

As in the aircraft gun the barrel recoils. The water jacket, however, is stationary since it is screwed on to the trunnion block. Thus packing glands must be provided near the breech and muzzle ends of the barrel to prevent water from escaping from the jacket where the barrel slides in and out of the jacket.

With these exceptions and a few changes in the accessories supplied, such as front and rear sights, the water cooled gun is identical with the aircraft gun.

# HOW THE GUN WORKS

*This smaller pamphlet was apparently for the operator of the machine gun.*

BROWNING MACHINE GUNS, CALIBER .50, M2

Aircraft (air-cooled) .......................... weight 64 lbs.

Anti-aircraft (water-cooled) ........... weight 121.5 lbs.

Tank and Field (air-cooled) ............. weight 81 lbs.

How the Gun Works, is a smaller pamphlet measuring 5"x3.5" and only 32 pages in its original form. This booklet teaches many of the same concepts as its larger brother in the earlier pages of this book. Likely if was used for training gunners and unit level armorers where as the larger book that leads off this reprint would likely have been used to train armorers at ordnance depots where full scaled maintenance would have been performed.

## INTRODUCTION

The Browning Machine Gun, caliber .50, is a highly efficient automatic weapon built to precision standards. It is produced with three types of barrels for various applications as shown to the left. The working parts, however, which automatically perform the numerous mechanical operations while the gun is firing, are the same in all three models. The following pages describe and illustrate "HOW THE GUN WORKS."

Each time a cartridge is fired, the mechanical action within the gun involves many moving parts. To gain a knowledge of the operation of these parts and their relationship to each other, the action has been separated into various phases. These are described in the following order:

1. FIRING

4. COCKING

2. RECOILING

5. AUTOMATIC FIRING

3. COUNTER-RECOILING

6. FEEDING

7. EXTRACTING AND EJECTING

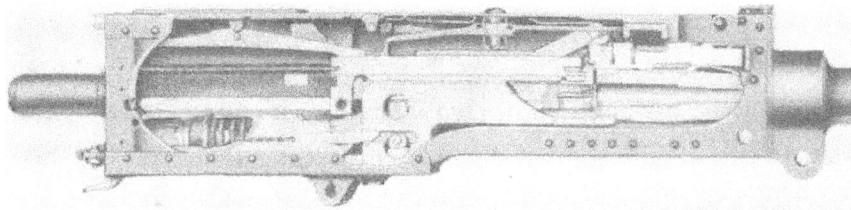

FIRING

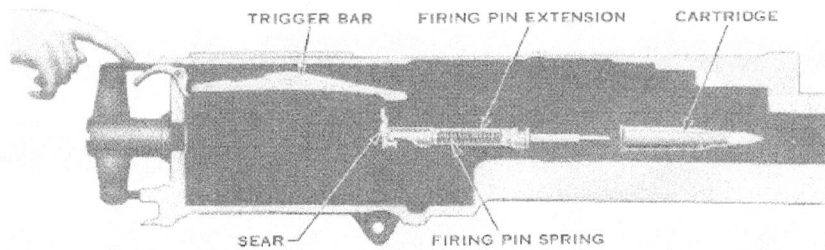

When the gun has been loaded and the firing pin spring has been cocked or compressed by hand, the firing mechanism is as shown. The gun is now ready to fire.

In this case a manual trigger and trigger bar are shown for firing, although for some applications the gun is fired by mechanical or electrical accessories.

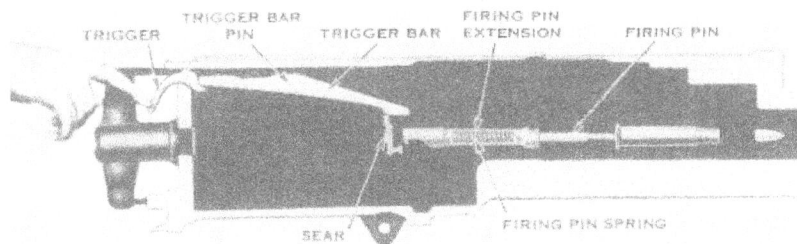

When the trigger is pressed it raises the back end of the trigger bar. The trigger bar pivots on the trigger bar pin, causing the front end to press down on the top of the sear. The sear is forced down until the notch in the sear is disengaged from the shoulder of the firing pin extension. The firing pin and firing pin extension are driven forward by the firing pin spring to fire the cartridge.

## RECOILING

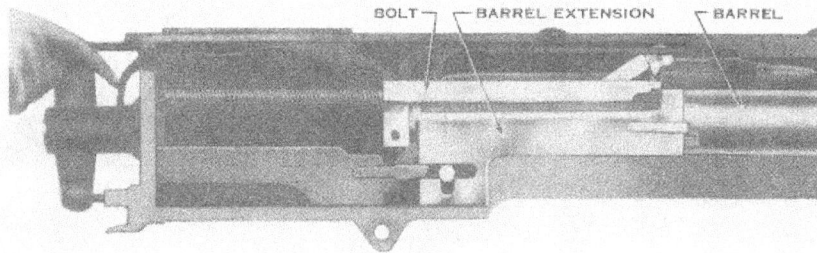

BOLT ── ┌─ BARREL EXTENSION    ┌─ BARREL

The complete cycle of the recoiling portion of the gun, which takes place as each cartridge is fired, consists of the recoil stroke when certain parts of the gun move rearward and the counter-recoil stroke when these same parts move forward. At the instant of firing, the barrel, barrel extension, and bolt, known as the recoiling portion, are in the forward position in the gun.

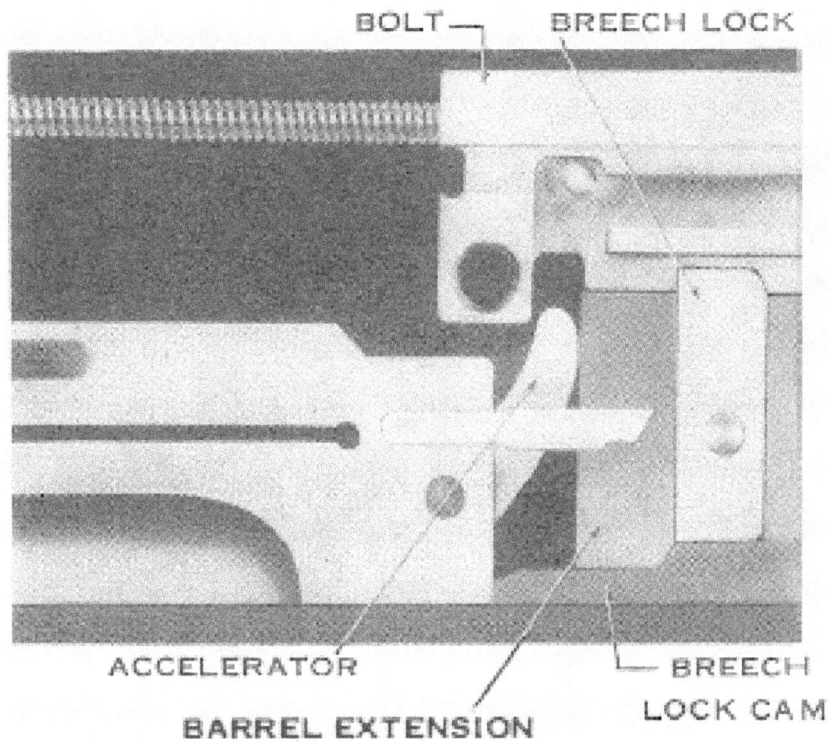

BOLT — BREECH LOCK

ACCELERATOR

BARREL EXTENSION

BREECH LOCK CAM

At this time the bolt is held securely against the base of the cartridge by the breech lock, which extends up from the barrel extension into a notch in the underside of the bolt.

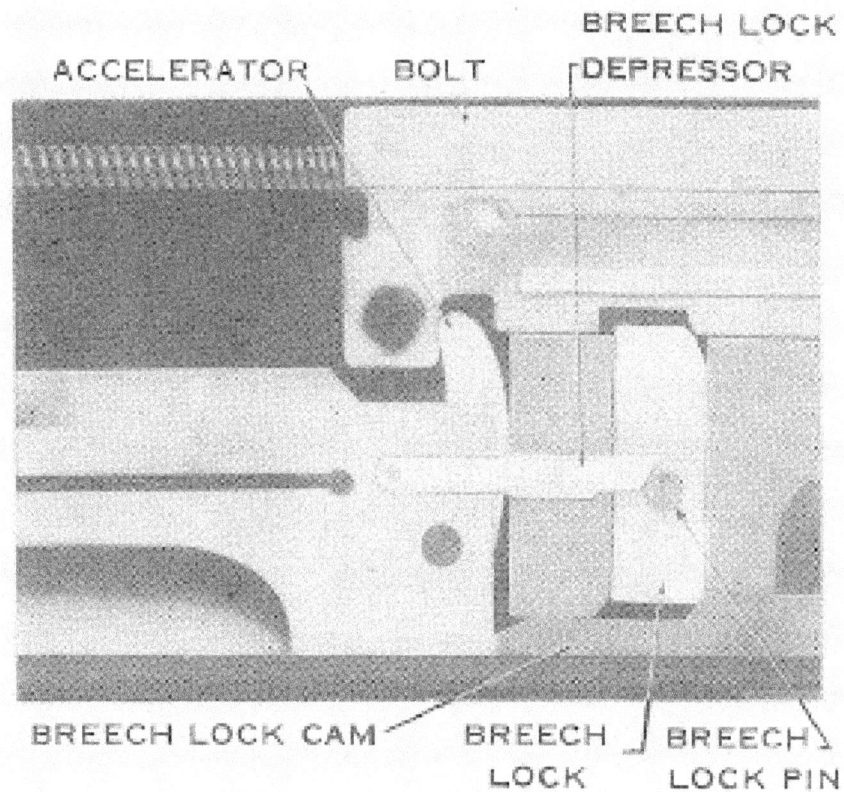

ACCELERATOR     BOLT     BREECH LOCK DEPRESSOR

BREECH LOCK CAM     BREECH LOCK     BREECH LOCK PIN

After the cartridge explodes and as the bullet travels out of the barrel, the force of recoil drives the recoiling portion rearward. During the first three- quarters inch of travel the breech lock is pushed back off the breech lock cam step. This permits the breech lock to be forced down out of the notch in the bolt by the breech lock depressors engaging the breech lock pin. This unlocks the bolt.

As the recoiling portion moves toward the rear the barrel extension rolls the accelerator rearward. The tip of the accelerator strikes the lower projection on the bolt and hastens or accelerates the bolt to the rear. (Note breech lock completely disengaged from bolt notch.)

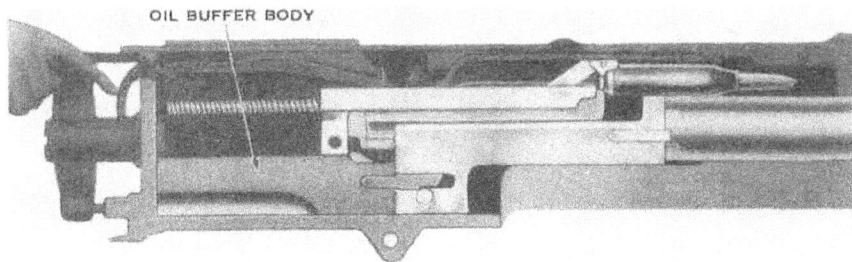

The barrel and barrel extension have a total rearward travel of one and one-eighth inches at which time they are completely stopped by the oil buffer body assembly.

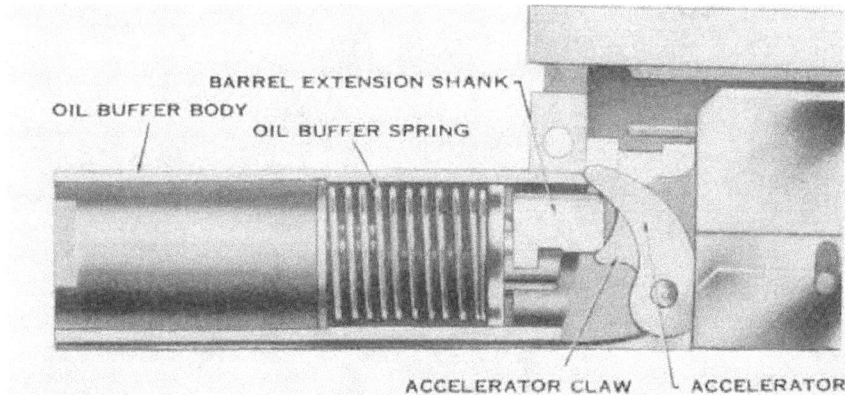

During this recoil of one and one-eighth inches the oil buffer spring is compressed in the oil buffer body by the barrel extension shank. The spring is locked in the compressed position by the claws of the accelerator which are moved against the shoulders of the barrel extension shank.

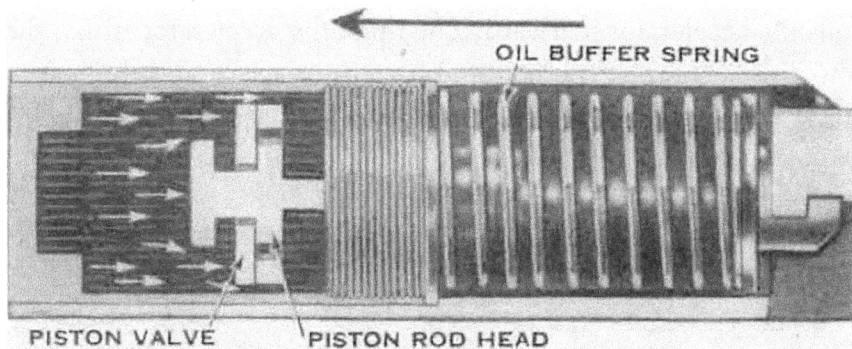

The oil buffer assists the oil buffer spring in bringing the barrel and barrel extension to rest during the recoil stroke. During the one and one-eighth inch of rearward travel the piston rod head is forced from the forward end of the oil buffer tube to the rear. The oil at the rear of the oil buffer tube under pressure of the piston escapes to the front side of the piston. Its only path is through restricted notches between the edge of the piston rod head and the oil buffer tube.

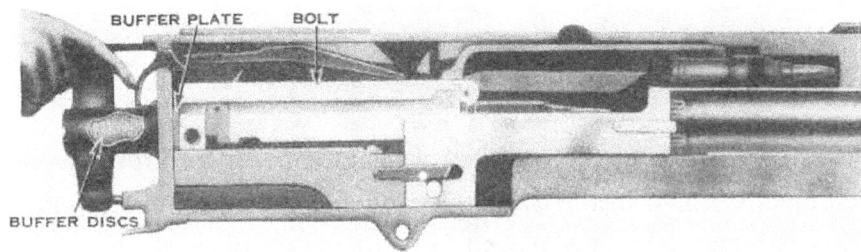

The bolt travels rearward tor a total of seven and one-eighth inches. During this travel the two, nested driving springs are compressed. The rearward stroke of the bolt is finally stopped as the bolt strikes the buffer plate. Thus, part of the recoil energy of the bolt is stored in the driving springs and the remainder is absorbed by the buffer discs in the back plate.

## COUNTER - RECOILING

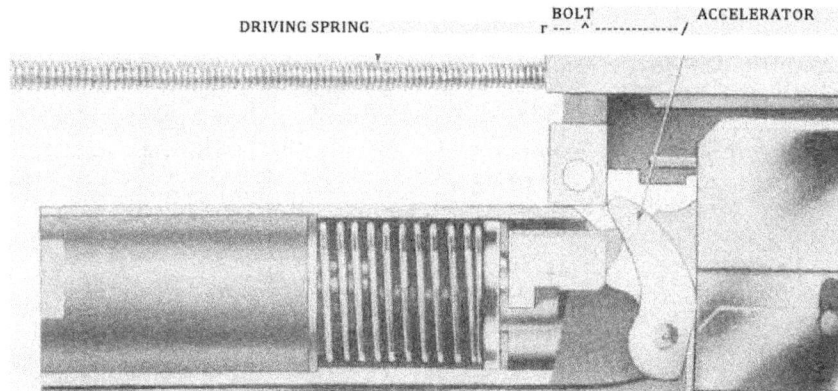

DRIVING SPRING               BOLT         ACCELERATOR

After completion of the recoil stroke the bolt is forced forward by the energy stored in the driving spring and the compressed buffer discs. When the bolt has moved forward about five inches the tip of the accelerator is struck by a projection on the bottom of the bolt. This rolls the accelerator forward.

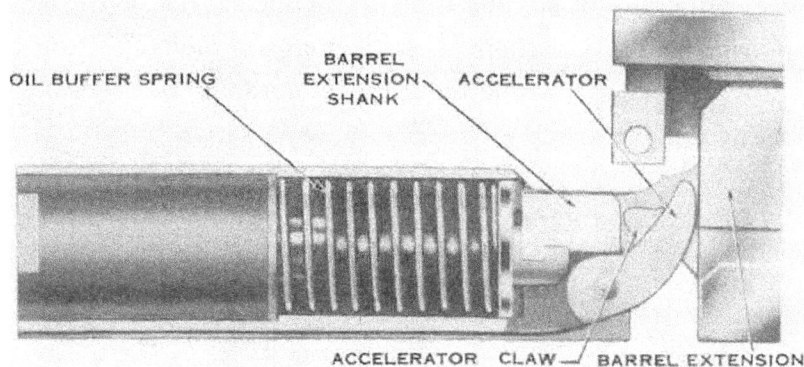

OIL BUFFER SPRING     BARREL EXTENSION SHANK    ACCELERATOR

ACCELERATOR CLAW    BARREL EXTENSION

As the accelerator rolls forward the accelerator claws are moved away from the shoulders of the barrel extension shank. This releases the oil buffer spring. The energy stored in the spring shoves the barrel extension and barrel forward.

PISTON VALVE-/            PISTON ROD HEAD

No restriction to motion is desired on the forward or counter-recoil stroke of the barrel and barrel extension; therefore, on the forward stroke additional openings for oil flow are provided in the piston rod head of the oil buffer assembly. The piston valve is forced away from the piston rod head as the parts move forward, uncovering these additional openings. This provides an additional path and permits oil to escape freely at the opening in the center of the piston valve as well as at the edge of the piston valve next to the tube wall.

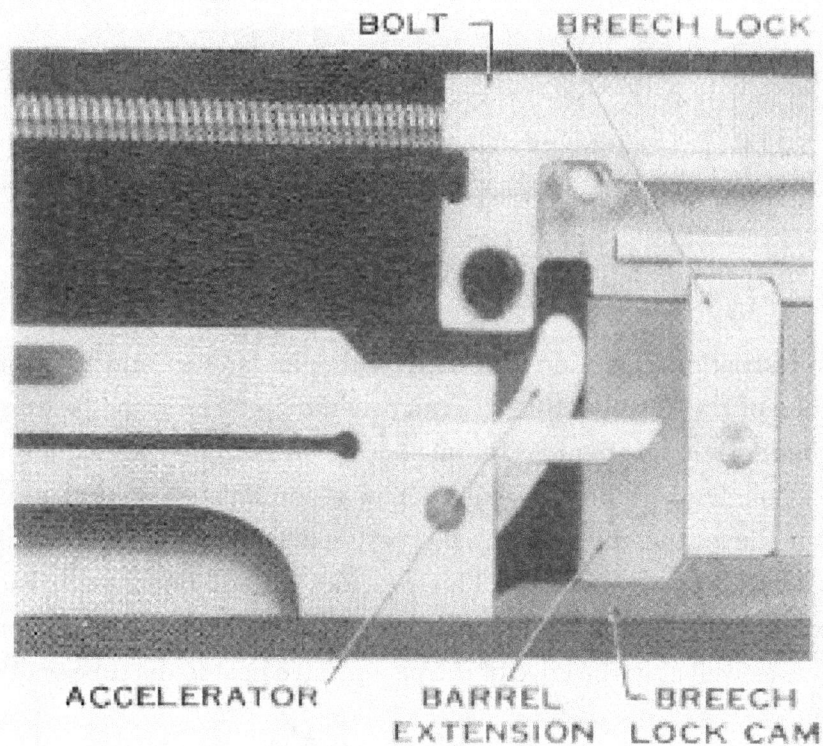

BOLT  BREECH LOCK

ACCELERATOR  BARREL  BREECH
EXTENSION  LOCK CAM

As the barrel extension moves forward the breech lock engages the breech lock cam and is forced upward. The bolt, which has been continuing its forward motion since striking the accelerator, has at this instant reached a position where the notch on the underside is directly above the breech lock, thus permitting the breech lock to engage the bolt. The bolt is thereby locked to the breech end of the barrel just before the recoiling portion reaches the firing position.

## COCKING

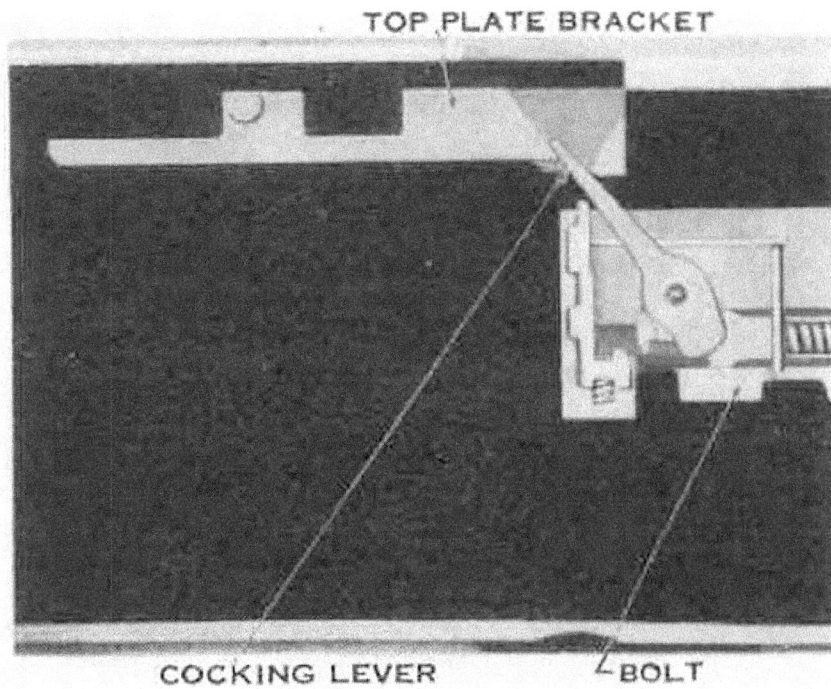

TOP PLATE BRACKET

COCKING LEVER BOLT

The act of cocking the gun is begun as the bolt starts to recoil immediately after firing. Thus the tip of the cocking lever, which is in the
V-slot in the top plate bracket, is forced forward.

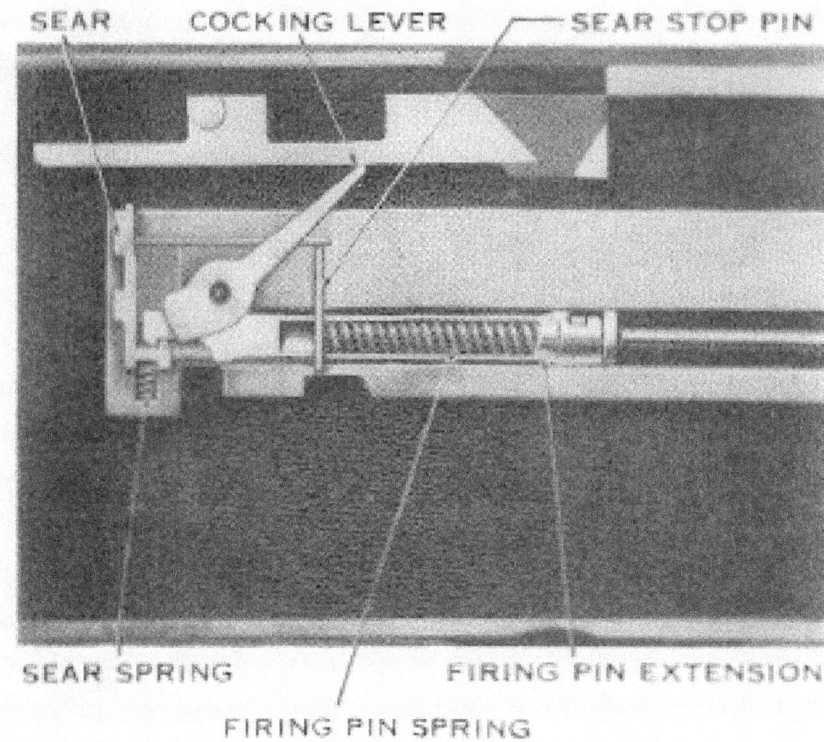

SEAR   COCKING LEVER   SEAR STOP PIN

SEAR SPRING   FIRING PIN EXTENSION

FIRING PIN SPRING

The cocking lever is pivoted so that the lower end forces the firing pin extension rearward. The firing pin spring is thus compressed against the sear stop pin. The shoulder at the back end of the firing pin extension is hooked over the notch at the bottom of the sear under pressure of the sear spring.

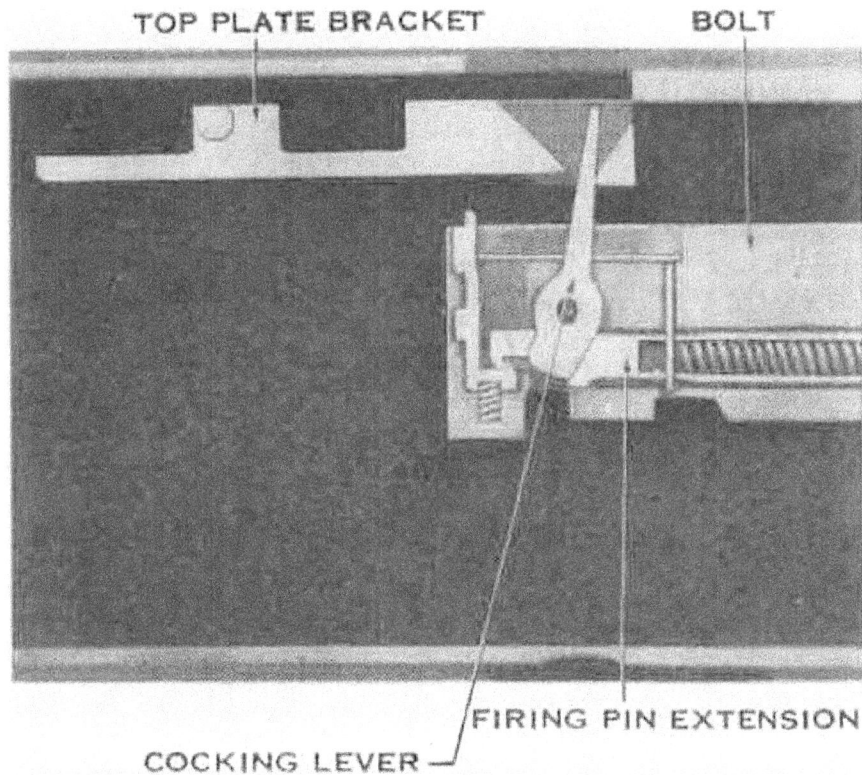

TOP PLATE BRACKET      BOLT

FIRING PIN EXTENSION

COCKING LEVER

During the forward motion of the bolt the tip of the cocking lever enters the V-slot of the top plate bracket. This action swings the bottom of the cocking lever out of the path of the firing pin extension; thus permitting the firing pin to snap forward to fire the cartridge.

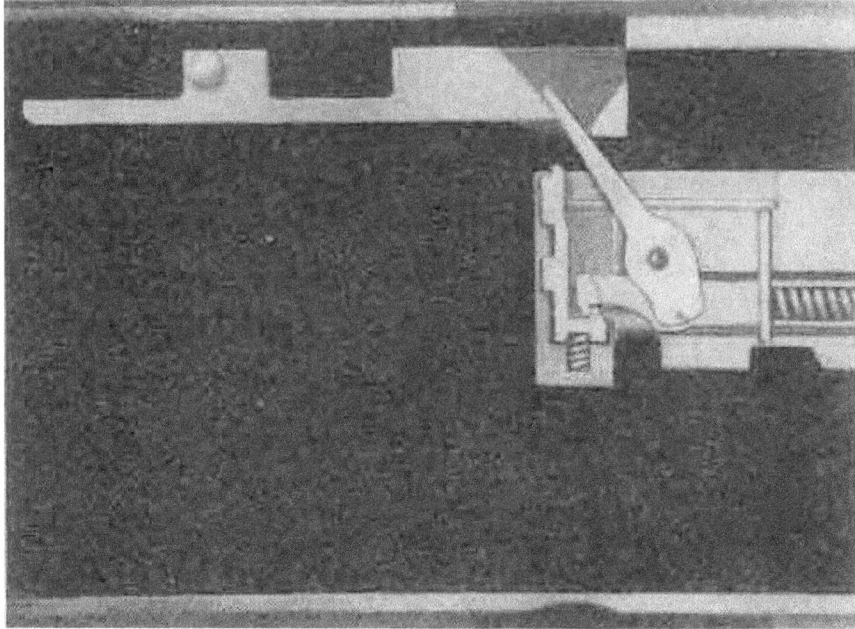

When the recoiling portion is almost in the forward position the gun is ready to fire. If no trigger action is given at this instant, the recoiling portion assumes its final forward position and the gun ceases to fire. The parts are now in the position shown on page 3 and the gun is again ready to fire.

## AUTOMATIC FIRING

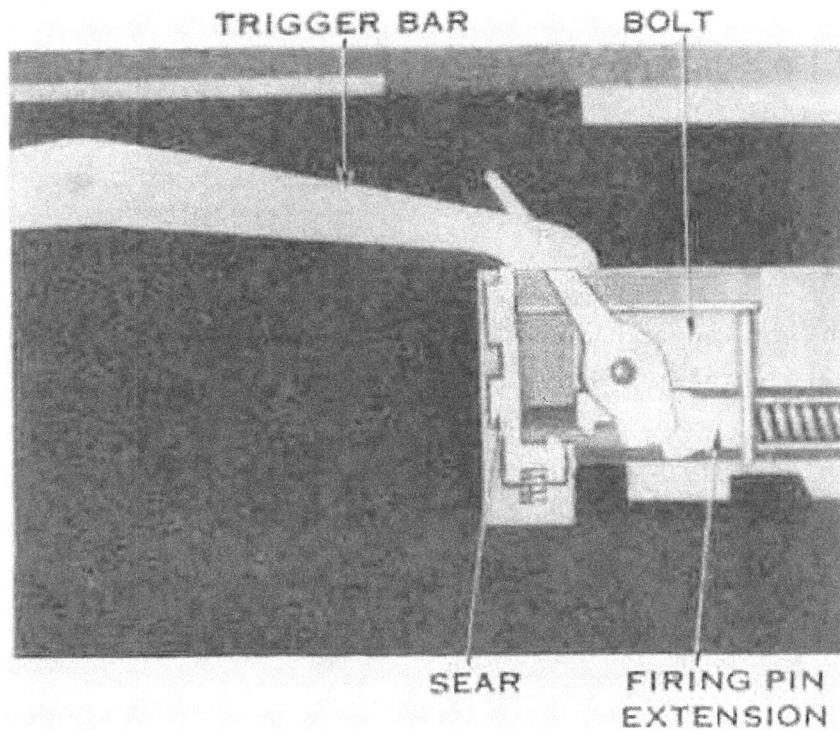

TRIGGER BAR      BOLT

SEAR      FIRING PIN EXTENSION

For automatic firing the trigger is pressed and held down. The sear is depressed as its tip is carried against the cam surface of the trigger bar by the forward movement of the bolt near the end of the counter-recoil stroke. The notch in the bottom of the sear releases the firing pin extension and the firing pin, thus automatically firing the next cartridge at the completion of the forward stroke. The gun fires automatically as long as trigger action is maintained and until the ammunition supply is exhausted.

## FEEDING

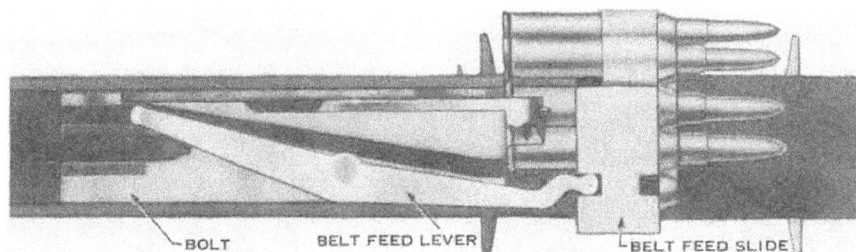

The belt feed mechanism is actuated by the bolt. When the bolt is in the forward position the belt feed slide is within the confines of the gun. This illustration shows the mechanism as from above with the cover removed. A stud at the rear of the belt feed lever is engaged in the diagonal groove or way in the top of the bolt.

As the bolt moves rearward during recoil the belt feed lever is pivoted. The forward end of the belt feed lever moves the belt feed slide out of the side of the gun and over the ammunition belt. Ammunition feed shown is from the left side of the gun. Feed from either side is possible with all caliber .50, M2 guns. (Note: On previous models, namely the M1921 and M1921 A, ammunition could be fed from only the left side.)

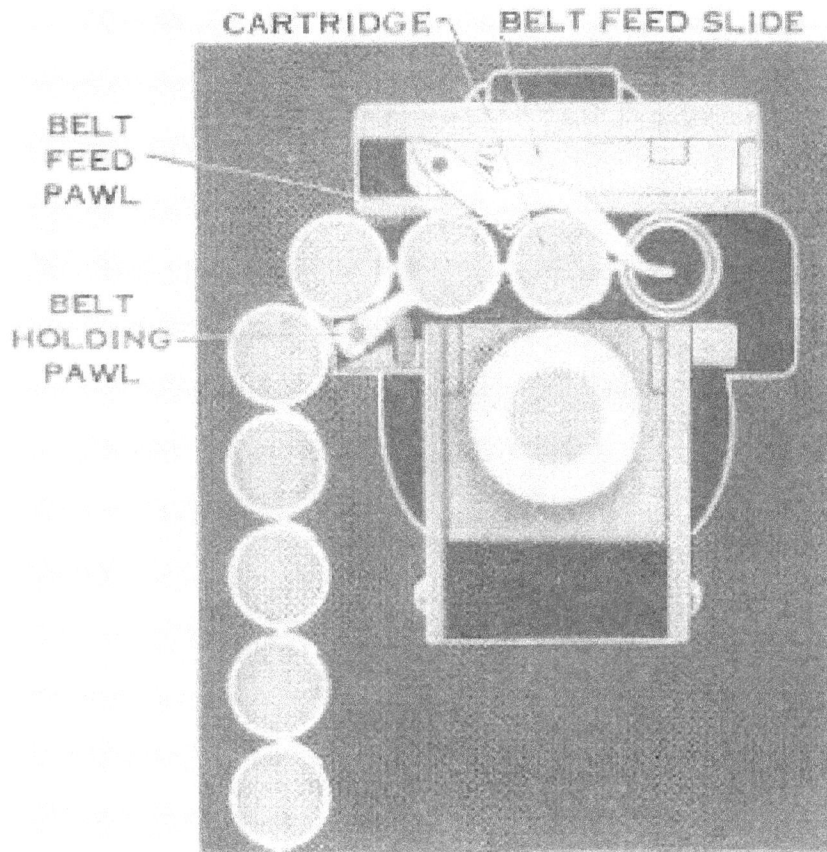

CARTRIDGE    BELT FEED SLIDE

BELT
FEED
PAWL

BELT
HOLDING
PAWL

The ammunition belt is pulled into the gun by the belt feed pawl which is attached to the belt feed slide. When the bolt is forward the belt feed pawl has positioned a cartridge directly above the chamber. The belt holding pawl is in a raised position to prevent the ammunition belt from falling out of the gun.

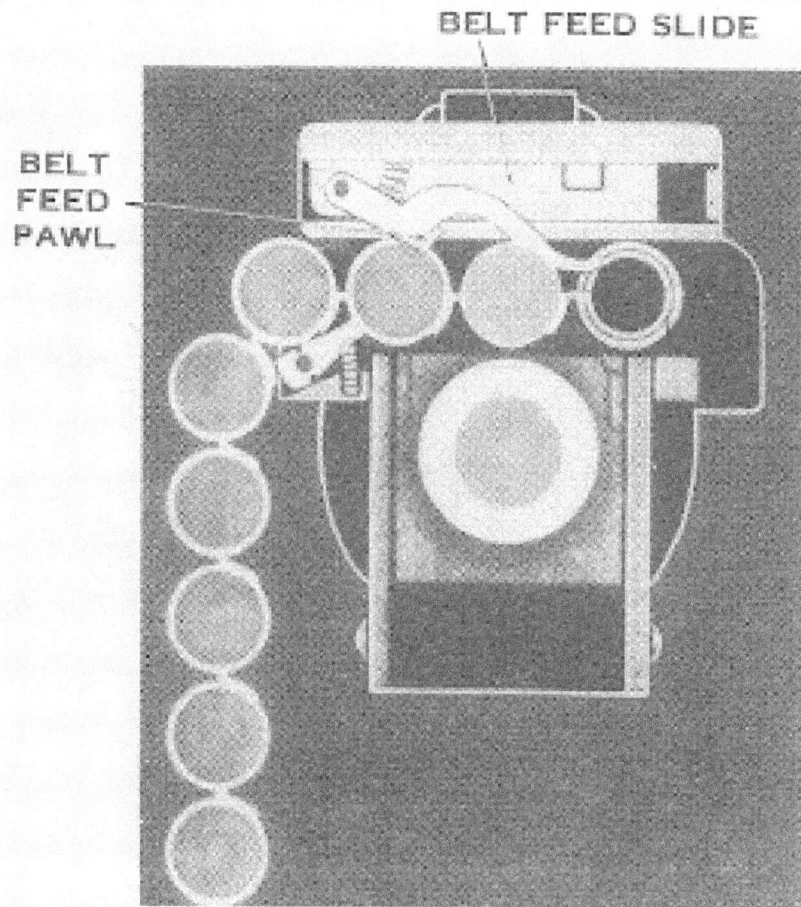

BELT FEED SLIDE

BELT
FEED
PAWL

As the bolt recoils the belt feed slide is moved out over the belt, and the belt feed pawl pivots so as to ride over the next cartridge.

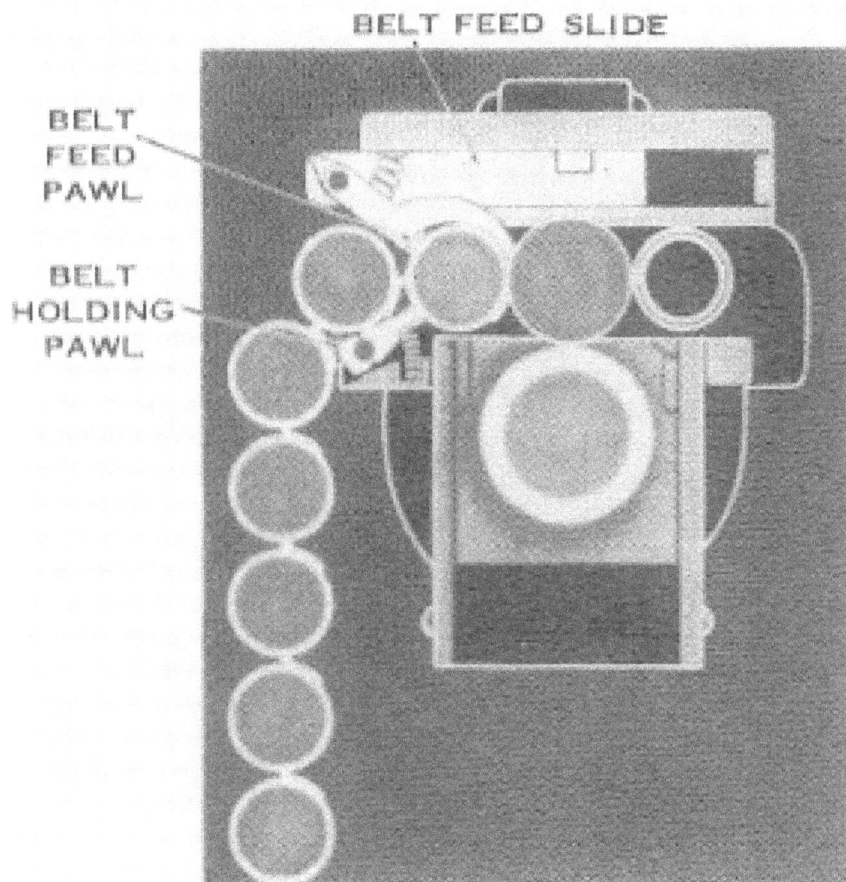

BELT FEED SLIDE

BELT
FEED
PAWL

BELT
HOLDING
PAWL.

At the end of the recoil stroke the travel of the belt feed slide is suffi-cient to permit the belt feed pawl to snap down behind the next cartridge in order to pull the belt into the gun.

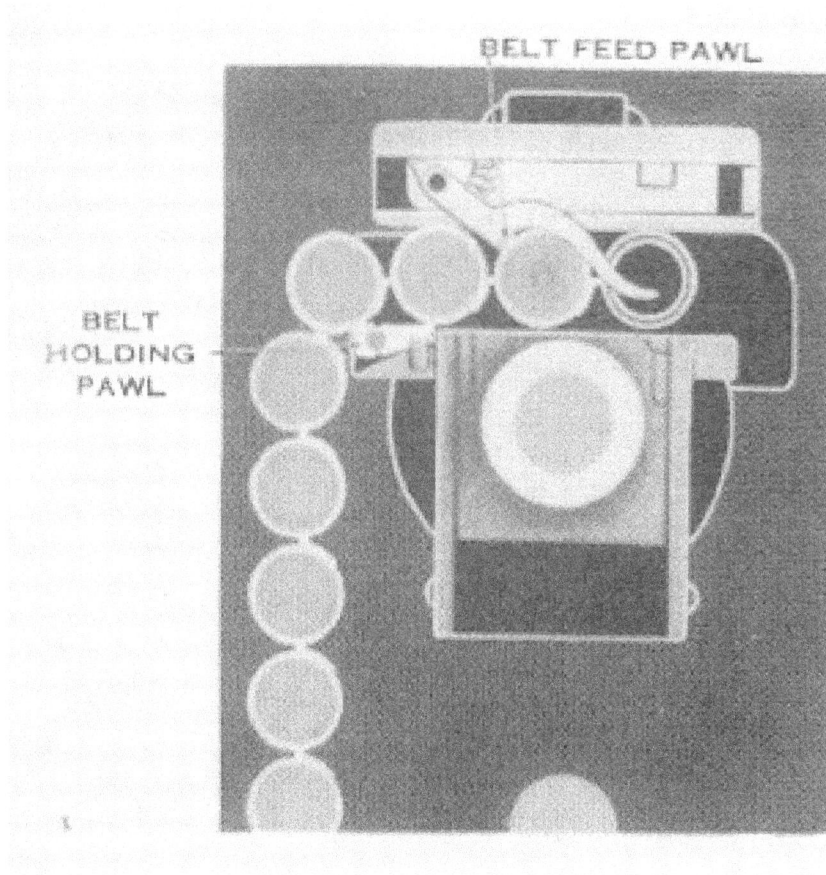

As the bolt moves forward on the counter-recoil stroke the belt is pulled into the gun by the belt feed pawl. The belt holding pawl is forced downward as a cartridge is pulled over it. When the forward stroke of the bolt is completed the belt holding pawl snaps up behind the next cartridge, as shown on page 24.

EXTRACTING AND EJECTING

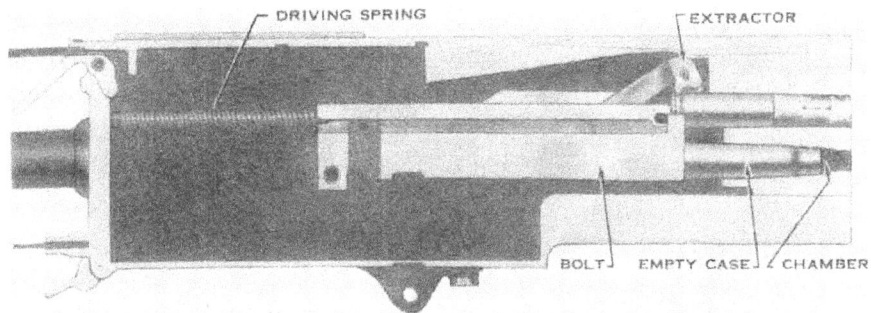

As recoil starts, a cartridge is drawn from the ammunition belt by the extractor. The empty case is withdrawn from the chamber by the T-slot in the front face of the bolt.

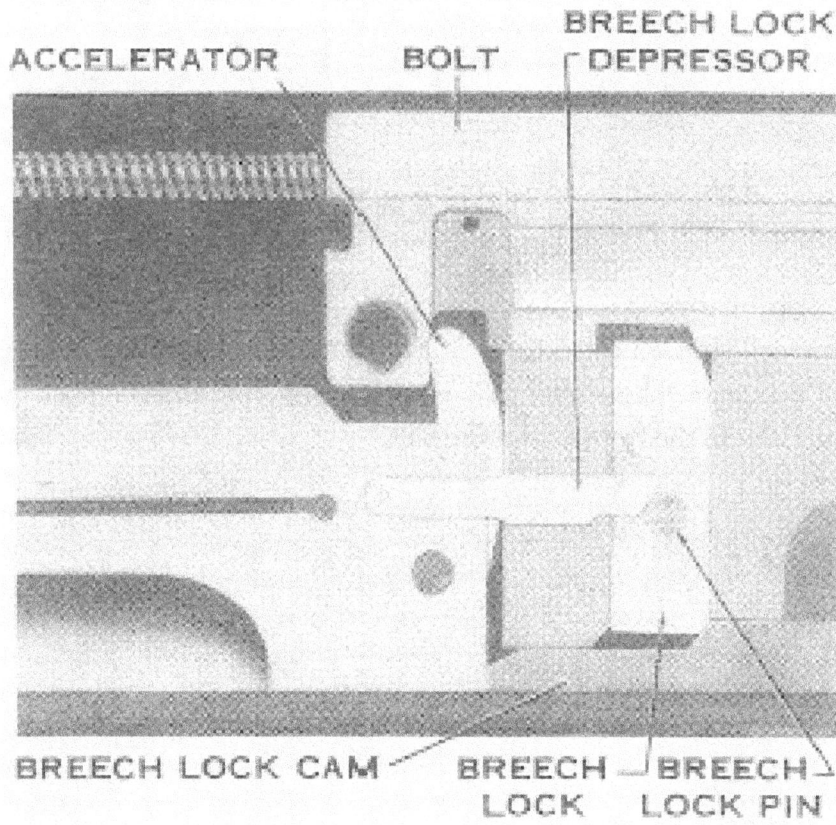

ACCELERATOR  BOLT  BREECH LOCK DEPRESSOR

BREECH LOCK CAM  BREECH LOCK  BREECH LOCK PIN

The empty case having been expanded by the force of explosion fits the chamber very snugly and the possibility exists of tearing the case if the withdrawal is too rapid. To prevent this and to insure slow initial withdrawal, the top front edge of the breech lock and front side of the notch in the bolt are beveled. Thus, as the breech lock is disengaged, the bolt moves away from the barrel and barrel extension in a gradual manner.

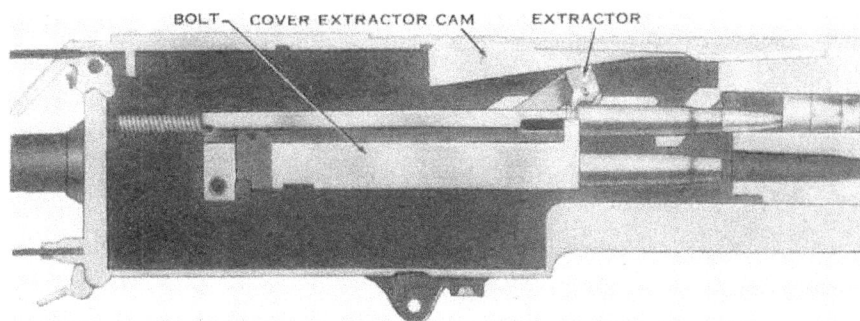

BOLT — COVER EXTRACTOR CAM — EXTRACTOR

As the bolt moves to the rear the cover extractor cam forces the extractor down, causing the cartridge to enter the T-slot in the bolt.

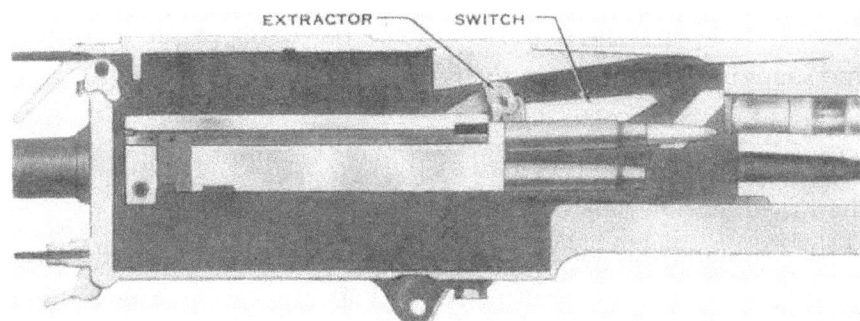

EXTRACTOR — SWITCH

As the extractor is forced down a lug on the side of the extractor rides against the top of the switch causing the switch to pivot downward at the rear. Near the end of the rearward movement of the bolt the lug on the extractor overrides the end of the switch, and the switch snaps up to its normal position.

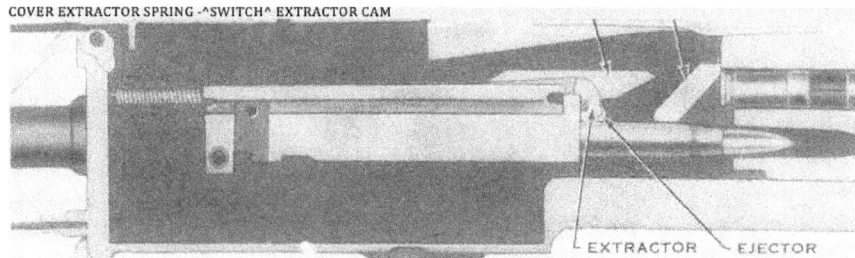

COVER EXTRACTOR SPRING -^SWITCH^ EXTRACTOR CAM

EXTRACTOR   EJECTOR

On counter-recoil the extractor is forced farther down by the extractor lug riding under the switch. The cartridge expels the empty case. The extractor stop pin in the bolt limits the travel of the extractor so that the cartridge, assisted by the ejector, enters the chamber. When the cartridge is nearly chambered the extractor rides up the extractor cam, compresses the cover extractor spring, and snaps into the groove in the next cartridge.

This booklet has been prepared by the AC Spark Plug Division and the Frigidaire Division of General Motors Corporation for the use of the Ordnance Department and all members of the United States armed forces who use, or are charged with the care of, the Caliber .50, Browning Machine Gun.

DETAILS OF BROWNING MACHINE GUN CALIBER .50, M2, AIRCRAFT BASIC

MANUFACTURED BY
AC SPARK PLUG DIVISION, General Motors Corporation, FLINT, MICHIGAN    FRIGIDAIRE DIVISION, General Motors Corporation, DAYTON, OHIO

For a free poster download of the Browning M2 Machine Gun go to: https://bit.ly/3tP5BvP

www.ingramcontent.com/pod-product-compliance
Lightning Source LLC
Chambersburg PA
CBHW080555090426
42735CB00016B/3248